THE PROFESSIONAL ENGINEERING CAREER DEVELOPMENT SERIES

ENGINEERING PROFESSION ADVISORY GROUP

Dr. Joseph M. Biedenbach, Manager Engineering Education Programs, RCA

Dean R. L. Bisplinghoff, School of Engineering, MIT

Dean Martin H. Bloom, College of Engineering, Polytechnic Institute of Brooklyn

Dean H. A. Bolz, College of Engineering, Ohio State University

Dr. John F. Conners, Director Manpower Development, Martin Marietta Corporation

Dean Elmer C. Easton, College of Engineering, Rutgers—The State University

Dr. Dolph G. Ebeling, General Electric Co., Schenectady, N. Y.

Mr. Ralph A. Emerson, Manager Educational Liaison, E. I. du Pont de Nemours & Co.

Dr. A. G. Hanson, President, The Georgia Institute of Technology

Dr. J. W. Haun, Vice President and Director of Engineering, General Mills, Inc.

Dean J. M. McKelvey, School of Engineering and Applied Science, Washington University

Dean W. W. Mullins, Carnegie Institute of Technology, Carnegie-Mellon University

Dean C. H. Norris, School of Engineering, University of Washington

Dean N. J. Palladino, College of Engineering, Pennsylvania State University

Mr. Frank G. Rizzardi, North American Rockwell, El Segundo, California

Dean Chauncy Starr, College of Engineering, University of California at Los Angeles

Dr. D. L. Thomsen, Jr., Director Engineering Education, IBM Corporation

Dean Robert E. Uhrig, College of Engineering, University of Florida

Professional Engineering Career Development Series

PROBABILITY for PRACTICING ENGINEERS

Henry L. Gray and Patrick L. Odell

BARNES & NOBLE, INC. NEW YORK
Publishers • Booksellers • Since 1873

L. C. Catalogue Card Number: 78-109531

SBN 389 00501 0

Distributed

In Canada
by the Ryerson Press, Toronto

In Australia and New Zealand
by Hicks, Smith & Sons Pty. Ltd.,
Sydney and Wellington

In the United Kingdom, Europe,
and South Africa
by Chapman & Hall Ltd., London

Printed in the United States of America

To Becky and Norma

Preface

This book is written for the practicing engineer with only a manipulative knowledge of elementary calculus. It is not a handbook, nor is it a theoretical development of probability. It is, however, an articulated discussion of the nature of probability and the manner in which probability can be of value to the engineer. Emphasis is placed on probability models and the manner in which they serve as proper mathematical models for a given physical phenomenon. Important concepts are set apart as definitions and theorems. Theorem proofs are only included when they are explanatory in nature. However, in most cases, the significance and implications of each definition and theorem are discussed and exemplified.

In Chapter I various definitions of probability are introduced and the manner in which these different notions complement each other is established. In addition, the important conditional probability function is discussed in detail, with special emphasis on its interpretation as an appropriate mathematical model for a given problem.

Chapters II and III are much the same nature as Chapter I, in that their primary purpose is to introduce to the engineer the concepts and terminology necessary for pursuing the applications of probability. In these chapters the notions of a random variable, a distribution function, a probability density function, and independence are considered.

Chapter IV is devoted to the application of particular probability density functions. Each density function is presented as an appropriate model for a wide range of physical problems, and a number of examples are given.

Chapter V introduces the concept of an estimator and establishes methods for estimating the parameters in the density functions of Chapter IV. In Chapters VI and VII these ideas are extended to stochastic processes: first to the problem of parameter estimation, and then to the problem of estimating a stochastic process.

Finally, the authors would like to express their gratitude to Kathy Dobbs and Linda Woolfolk for their time in typing the manuscript.

Contents

I
Basic Concepts

1.1 INTRODUCTION

There are a number of ways to define probability, although there is no "best" way, since "best" must be interpreted in terms of individual interest and needs. This book is primarily for the practicing engineer, so the approach will be a mixture of the relative frequency, a priori, and axiomatic definitions. The relative frequency and a priori definitions will be used primarily to make our study heuristic in nature. In this way it will be possible to intuit or at least develop a "feel" for some of the more important concepts. As we shall soon see, however, these approaches are not adequate and at times we will be forced to call on the axiomatic definition. However, even then it will be clear that the basic axioms are products of the relative frequency and/or a priori definitions. We will now describe briefly these three concepts.

1.2 A PRIORI PROBABILITY

Ask almost any housewife the probability of obtaining a head on a single toss of a coin and she will answer "one half." Ask her why it is one half and she will reply that it must land heads or tails. This simple example illustrates the crux of the a priori definition, the appealing aspects of the definition, and its inadequacy. To be more specific let us make the following definition.

Definition 1 If the total number of possible outcomes of an experiment is N and if in N_1 of these outcomes the event A occurs, then we define the a priori probability, $P[A]$ (read probability that the event A occurs), by

$$P[A] = \frac{N_1}{N}. \tag{1}$$

In the above, and henceforth, the word experiment will refer to a process which is hypothetically repeatable under identical conditions. We will restrict questions of probability to such experiments. Thus the question, "What is the probability of obtaining a head in one toss of a coin?" will be considered appropriate, while the question, "What is the probability my wife loves me?" will not.

If we define the total number of possible outcomes when a coin is tossed as 2 (that is, heads or tails), then we see that the a priori probability of a head is given by

$$P[\text{Head}] = \tfrac{1}{2},$$

as claimed above, and hence the definition has the advantage of simplicity as well as appealing to intuition. Note that it is necessary to define the possible outcomes of the experiment. For example, if we allow heads, tails, stand on the edge, or lost coin as possible outcomes, then we have four possible outcomes and the a priori probability is given by

$$P[\text{Head}] = \tfrac{1}{4}.$$

This does not seem reasonable even if we did want to allow stand on the edge and lost coin as possible outcomes, since for the normal coin we would not consider these outcomes as having the same likelihood as heads or tails. This immediately points to one of the difficulties in using (1): the fact that (1) is not physically meaningful unless the outcomes of the experiment are equally likely. Moreover, since by equally likely we mean equally probable, we see that there is no way to correct this difficulty in our definition since in trying to correct it we make the definition circular. Even if we were willing to leave the words *equally likely* undefined, we see that the definition would be of use in only a limited class of problems. The difficulties present in using (1) have possibly been oversimplified in this discussion, and we should caution the reader somewhat more. That is, it is granted that there are cases in which the definition of a priori probability will suffice. However, in the problem discussed above the question of equally likely is intuitively satisfied for the experiment, in which heads and tails form the totality of outcomes. For more complicated problems this is often not the case, and the question of equally likely can be a difficult one and hence the use of (1) may be questionable.

1.3 RELATIVE FREQUENCY PROBABILITY

A definition which has the same intuitive advantage as (1), Section 1.2, but which does not suffer from the problem of necessitating equally likely outcomes is one based on relative frequency. That is, if an experiment is repeated a large number of times n and the event A occurs n_1 of these times, then we consider n_1/n as an approximation to the probability of A and we define $P[A]$ by

$$P[A] = \lim_{n\to\infty} \frac{n_1}{n}. \tag{1}$$

Thus in the coin problem no difficulty is incurred if we allow standing on the edge or losing the coin as possible outcomes. The definition is attractive to engineers and physicists because it describes exactly the phenomenon as it has been observed. In fact, the definition is adhered to by most people naturally. Most will agree that when we say the probability of a head is $\frac{1}{2}$, we mean that on a large number of tosses we would expect to get heads about 50 percent of the time. Hence the definition indicated by (1) seems the most natural one.

Although the relative frequency approach to probability is essential for applying probability in the physical world, it is not satisfactory as the basis for the mathematical structure needed to interpret the complex problems of modern probability. However, this does not prevent us from using it as our guide whenever possible. In fact, we will find both it and the notion of a priori probability invaluable tools in the study of probability.

1.4 AXIOMATIC PROBABILITY

Although we have labeled this section "axiomatic probability" we will not go very far in that direction. In fact, we use this section primarily for the purpose of defining some needed terms and for indicating the manner in which the axiomatic definition removes the difficulties encountered in the previously discussed approaches. We also point out that even in the axiomatic development one finds a need for the notions of probability already discussed.

Connected with any experiment is a set of all possible outcomes of the experiment. As mentioned before, the definition of this set is the prerogative of the experimenter. We shall hold this definition as self-evident and not consider it further. Actually there are often some subtle difficulties involved in defining the sample space. It would be confusing to try to discuss them here and so we will not attempt it. We will, however, touch on these problems when we discuss the notion of conditional probability. In any case, when we consider a set of all possible outcomes of an experiment we will refer to it as the *sample space* or fundamental probability set or space. We will usually denote this set by Ω. Elements of Ω (that is, individual outcomes of an experiment) will be denoted by ω. Any collection of elements of Ω will be called an *event*. Strictly speaking, this definition of an event is mathematically incorrect and would lead to certain unavoidable mathematical difficulties upon close scrutiny. However, these difficulties are pathological and not of a practical nature, and they will cause the engineer no problems. We will say an event E *occurs* if any element (that is, experimental outcome) in E occurs. An impossible outcome of an experiment, for example, three heads on two tosses of a coin, will be referred to as an *impossible event* and will be denoted by ϕ. We will consider the impossible event as a legitimate event under the definition above by interpreting it as the empty collection of elements of Ω. As an event the set Ω will be called the *sure event*.

Two events E_1 and E_2 will be said to be *mutually exclusive* or *disjoint* if they have no elements in common, that is, if the occurrence of E_1 excludes the simultaneous occurrence of E_2 and vice versa. We denote this by $E_1 E_2 = \phi$. The notation

$E_1 E_2$ is read "the event composed of E_1 and E_2." By the symbol $E_1 \cup E_2$ (read E_1 or E_2) we will mean the event composed of the outcomes in E_1 or E_2 or both. If $E_1 E_2 = \phi$ we will write $E_1 + E_2$ rather than $E_1 \cup E_2$. The above notation is extended jointly to more than two events. For example, the n events $E_1, E_2, \ldots, E_n$ are said to be mutually exclusive if $E_i E_j = \phi$ when $i \neq j$. We leave further extensions to the reader, as they are obvious. We summarize below.

Ω = fundamental probability set or space
ω = element of Ω
ϕ = impossible event
$E_1 E_2$ = event composed of the elements in E_1 and E_2
$E_1 E_2 = \phi$ = mutually exclusive events
$E_1 \cup E_2$ = event composed of the elements in E_1 or E_2 or both
$E_1 + E_2$ = same as $E_1 \cup E_2$ but in this case $E_1 E_2 = \phi$

Consider the experiment of tossing a coin twice and observing the property of heads and tails. In this case we define the probability space Ω by the set of possible outcomes {HH,HT,TH,TT}. To illustrate the above ideas, we will now list some events of this experiment.

(a) Ω = {HH,HT,TH,TT}.
(b) E_1 = {HH,HT,TH} = event of at least one head.
(c) E_2 = {HH} = event heads on the first toss and heads on the second.
(d) E_3 = {HT} = event heads on the first toss and tails on the second.
(e) $E_2 + E_3$ = event of heads on the first toss and heads or tails on the second.
(f) $E_1 E_3 = E_3$. This is true since clearly HT is the only element common to both.
(g) If E_4 = {TT}, then $E_4 E_3 = \phi$, since one cannot obtain the outcomes TT and HT simultaneously. Note that E_4 and E_3 are therefore mutually exclusive events.
(h) $E_4 + E_1 = \Omega$.

We are now in a position to state the axiomatic definition of probability.

Definition 1 A probability function P is a function which assigns to every event E a real number $P[E]$ called the probability of E or the probability E occurs. The function P is assumed to satisfy the following axioms.

(a) $P[E] \geq 0$ for every event E.

(b) If $E_1, E_2, \ldots, E_n, \ldots$ are mutually exclusive events, then

$$P\left[\sum_{i=1}^{\infty} E_i\right] = \sum_{i=1}^{\infty} P[E_i].$$

(c) $P[\Omega] = 1$.

For the proper assumptions on the subsets of Ω which we have referred to as events, axioms (a), (b), and (c) are sufficient to develop the entire mathematical structure of probability. Throughout the remaining part of the book the words probability function will refer to a function satisfying Definition 1.

Two comments should be made in regard to this definition. First, it is clear that the previously discussed probability functions satisfy the conditions of this definition and hence are legitimate probability functions. In fact, one might say that it was the notion of a priori probability which formulated the basis for the selection of axioms (a), (b), and (c). However, in the axiomatic case we are not held to such a restricted notion of probability as in the previous definitions. This flexibility, although desirable, leads us to the second comment: It should be noticed that the axiomatic definition allows any function which satisfies the conditions of the definition to be a probability function. That is, no requirement other than those stated in the definition is placed on the way in which we assign a number to $P[E]$. To be more specific, consider again the case where a coin is tossed. For all practical purposes the relative frequency and a priori definitions would yield

$$P[\text{Head}] = \tfrac{1}{2}$$

for a fair coin. On the other hand, the axiomatic definition does not say what number we should assign $P[\text{Head}]$. Thus, we could define P by $P[\text{Head}] = \frac{1}{8}$ and $P[\text{Tail}] = \frac{7}{8}$ or $P[\text{Head}] = \frac{1}{2} = P[\text{Tail}]$. Hence we see that this definition does not solve the dilemma of the engineer, that is, the problem of assigning a number to $P[E]$.

This problem is solved by the adoption of probability models which we will discuss later. In each case, these models will be motivated by our previous definitions of probability, i.e., by a priori or relative frequency considerations. Henceforth, when the assumption of equally likely outcomes is reasonable and the experiment has only a finite number of possible outcomes, we will define the probability function P in the axiomatic definitions by (1) of Section 1.2; that is,

$$P[E_1] = \frac{N_1}{N}.$$

In regard to probability functions we list the following theorem.

Theorem 1 If P is a probability function, then for any events E_1 and E_2 we have the following:

(a) $P[E_1 + E_2] = P[E_1] + P[E_2]$.
(b) $P[E_1 \cup E_2] = P[E_1] + P[E_2] - P[E_1 E_2]$.
(c) $P[\phi] = 0$.
(d) $P[E] \leq 1$ for every event E.
(e) If E_1^c is the set of all possible outcomes not in E_1, then $P[E_1] = 1 - P[E_1^c]$.

Proof: We will prove the above theorem only for the case of a finite number of equally likely outcomes. We do this only in order to demonstrate the plausibility of the theorem. To show (b) consider Figure 1.1. In the figure we have represented the

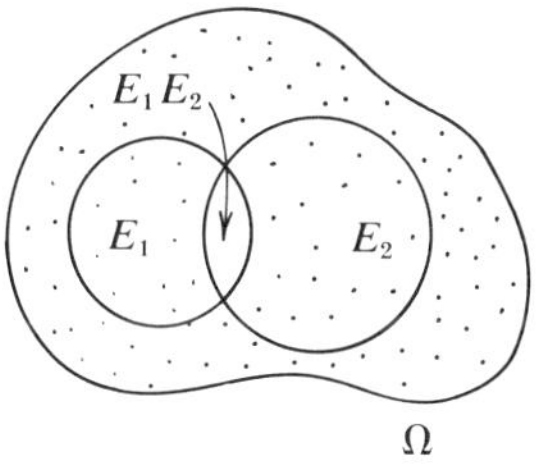

Figure 1.1 Venn Diagram of the Events E_1, E_2, and Ω

events E_1 and E_2 by circles and the outcomes in these events by dots. Now if the number of dots in E_1 is N_1 and the number of dots in E_2 is N_2, then clearly the number of dots in $E_1 \cup E_2$ (that is, E_1 or E_2) is $N_1 + N_2 - N_3$, where N_3 is the number of dots in E_1E_2. Therefore, if the total number of possible outcomes in Ω is N we have

$$P[E_1 \cup E_2] = \frac{N_1 + N_2 - N_3}{N} = \frac{N_1}{N} + \frac{N_2}{N} - \frac{N_3}{N}$$
$$= P[E_1] + P[E_2] - P[E_1 E_2].$$

Now if $N_3 = 0$, that is, E_1 and E_2 have no outcomes in common, then $E_1 \cup E_2 = E_1 + E_2$ and we have

$$P[E_1 + E_2] = P[E_1] + P[E_2].$$

To show (c) and (d) we simply note that $0/N = 0$, and $N_1/N \leq 1$ for any N_1 such that $0 \leq N_1 \leq N$. Finally, to show (e) we note if N_1 of the total possible outcomes are in E_1, then there are $N - N_1$ possible outcomes in E_1^c. Thus

$$P[E_1] + P[E_1^c] = \frac{N_1}{N} + \frac{N - N_1}{N} = \frac{N}{N} = 1,$$

the desired result. Note that since $E_1 + E_1^c = \Omega$, the result follows in general from axiom (c).

Some simple examples are given now to illustrate these results.

Example 1. Suppose that golf balls are produced by a process in which it has been established that approximately 1 percent of the balls will be heavy and $\frac{1}{2}$ percent will have cuts on the cover. In addition, assume that all golf balls are checked for these imperfections before packaging and rejected as defective if they possess either defect. What is the probability that a golf ball selected at random will be rejected? Now let

$$E_1 = \text{the event of a ball being heavy}$$

and

$$E_2 = \text{the event of a ball being cut.}$$

Then E_1 occurs if the ball is heavy and E_2 occurs if it is cut. Moreover, $E_1 E_2 \neq \phi$, since a ball may be both heavy and cut.

Thus

$$P[E_1 \cup E_2] = P[E_1] + P[E_2] - P[E_1E_2]$$
$$= .01 + .005 - P[E_1E_2].$$

Now, if we can determine $P[E_1E_2]$ the problem is solved. If we reason that the events E_1 and E_2 are in no way related but are independent phenomena, then based on a priori probability (this will be explained more completely in Chapter IV) we would define $P[E_1E_2] = P[E_1]P[E_2]$. Thus we have

$$P[E_1 \cup E_2] = .01 + .005 - .00005 = .01045.$$

Example 2. Consider the same process as discussed in Example 1, but assume that the possibility of a ball being cut is negligible and that the rejection criterion is heavy or light rather than heavy or cut. In this case let

$$E_1 = \text{the event of a ball being heavy}$$

and

$$E_2 = \text{the event of a ball being light.}$$

Then certainly the events are mutually exclusive, since no ball can be both heavy and light. If it is known that balls are as likely to be light as heavy, then we have

$$P[E_1 \cup E_2] = P[E_1 + E_2] = P[E_1] + P[E_2] = .02.$$

1.5 CONDITIONAL PROBABILITY

In this section we will discuss a very important probability function referred to as the conditional probability function.

Returning to Example 1 of Section 1.4, suppose we would like to know the probability that a heavy ball will also have a cut. If we use the relative frequency approach, then the probability that a ball is cut given that it is heavy (denoted $P[E_2 \mid E_1]$) is given by

$$P[E_2 \mid E_1] = \lim_{N_1 \to \infty} \frac{N_3}{N_1},$$

where N_1 is the number of balls which are observed heavy and N_3 is the number of balls which are observed to be heavy and cut. Now since $N_1 = (.01)\,N$ we see that $N_1 \to \infty$ if and only if $N \to \infty$. Thus

$$\lim_{N_1 \to \infty} \frac{N_3}{N_1} = \lim_{N \to \infty} \frac{\frac{N_3}{N}}{\frac{N_1}{N}}$$

$$= \frac{\lim_{N \to \infty} \frac{N_3}{N}}{\lim_{N \to \infty} \frac{N_1}{N}} = \frac{P[E_1 E_2]}{P[E_1]},$$

provided these limits exist and the one in the denominator is not zero. Thus

$$P[E_2 \mid E_1] = \frac{P[E_1 E_2]}{P[E_1]}.$$

This leads us to the following definition.

Definition 1 If E_1 and E_2 are any events and $P[E_1] > 0$, then we define the conditional probability of the event E_2 given the event E_1 has occurred, $P[E_2 \mid E_1]$, by

$$P[E_2 \mid E_1] = \frac{P[E_1 E_2]}{P[E_1]}. \tag{1}$$

It can be shown that the conditional probability function defined by (1) is indeed a probability function; i.e., it satisfies the conditions of Definition 1, Section 1.4.

The following theorems are quite useful and the reader should study them closely.

Theorem 1 If $E_1, E_2, \ldots, E_n$ are events and $P[E_1 E_2 \cdots E_n] > 0$, then

$$P[E_1 E_2 \cdots E_n] = P[E_1]\,P[E_2 \mid E_1]\,P[E_3 \mid E_1 E_2] \cdots \cdots P[E_n \mid E_1 \cdots E_{n-1}]. \tag{2}$$

Theorem 2 If $E_1, E_2, \ldots, E_n$ are mutually exclusive events such that

$$\sum_{i=1}^{n} P[E_i] = 1$$

and if E is any event such that $P[E] > 0$, then

$$P[E_k \mid E] = \frac{P[E \mid E_k]\,P[E_k]}{\sum_{i=1}^{n} P[E \mid E_i]\,P[E_i]}. \tag{3}$$

Equation (3) is called *Bayes's rule*. It will become more meaningful upon consideration of the following examples.

Example 1. A box contains R red balls and B black balls. A ball is drawn from the box and then returned to the box along with m more balls of the same color. What is the probability that one will draw a red ball three times in three successive draws?

The solution to this problem is quite simple if one makes use of (2). Let E_i be the event of drawing a red ball on the ith draw. Then

$$\begin{aligned} P[E_1 E_2 E_3] &= P[E_1]\,P[E_2 \mid E_1]\,P[E_3 \mid E_1 E_2] \\ &= \frac{R}{R+B}\,\frac{R+m}{R+B+m}\,\frac{R+2m}{R+B+2m}. \end{aligned}$$

Example 2. In a factory, machine a produces 20 percent of the days output, machine b produces 50 percent, and machine c produces the remaining 30 percent. One percent of the output of machine a, 2 percent of the output of machine b, and 3 percent of the output of machine c is defective. In a day's run the three machines put out 100,000 items. An item selected at random is found to be defective. What is the probability that it was produced by machine c?

Solution: Let d denote defective and A, B, and C be the event of an item coming from machines a, b, and c, respectively. We desire to find $P[C \mid d]$. The problem at first glance may seem quite difficult, but it can be solved very easily by Bayes's rule. (This is often the case when one has the situation where $P[E_1 \mid E_2]$ is difficult to calculate directly yet

$P[E_2 \mid E_1]$ is quite simple to obtain.) By (3),

$$P[C \mid d] = \frac{P[Cd]}{P[d]}$$

$$= \frac{P[d \mid C]\,P[C]}{P[d \mid A]\,P[A] + P[d \mid B]\,P[B] + P[d \mid C]\,P[C]}.$$

But

$$\begin{aligned} P[d \mid C] &= .03 \\ P[d \mid A] &= .01 \\ P[d \mid B] &= .02 \\ P[A] &= .20 \\ P[B] &= .50 \end{aligned}$$

and

$$P[C] = .30.$$

Consequently,

$$P[C \mid d] = \frac{.009}{.001 + .01 + .009} = \frac{9}{20}.$$

Some comment should be made in regard to the application or appropriateness of (1) in a physical setting. Consider the following example.

Example 3. A student is taking a true–false quiz which contains only two questions. He goes to the professor's desk to ask a question and sees that one of the questions is marked true, but he does not see which question is true. Upon returning to his seat he marks both questions true. Assuming the student has no other knowledge as to the answers to the questions, what is the probability he will get both questions correct?

Let

$$E_1 = \text{the event that both questions are true}$$

and

$$E_2 = \text{the event that at least one is true.}$$

It seems logical that we are simply asking for $P[E_1 \mid E_2]$. If this is the case, then we have

$$P[E_1 \mid E_2] = \frac{P[E_1 E_2]}{P[E_2]}.$$

If we assume true and false are equally likely, then

$$P[E_1E_2] = P[E_2 \mid E_1]\,P[E_1] = P[E_1] = \tfrac{1}{4}$$

and

$$P[E_2] = \tfrac{3}{4},$$

since

$$E_2 = \{\text{TT,TF,FT}\}.$$

Thus

$$P[E_1 \mid E_2] = \frac{\frac{1}{4}}{\frac{3}{4}} = \tfrac{1}{3}.$$

Another way to get this same result is as follows. Prior to going to the professor's desk, the probability space, Ω, of possible answers to the quiz was

$$\Omega = \{\text{TT,TF,FT, FF}\}. \tag{4}$$

One and only one of these combinations of answers is correct. Under the assumption that all are equally likely and the student is simply guessing we must assign a probability of $\frac{1}{4}$ to each of the possible outcomes, and hence $P[E_1] = \frac{1}{4}$. Upon going to the professor's desk, the student has gained additional information and the probability space is no longer given by (4) but by

$$\Omega_1 = \{\text{TT,TF,FT}\}. \tag{5}$$

Now, since no additional information is available and one and only one of the combinations in (5) is correct, each being equally likely, we see that again $P[E_1 \mid E_2] = \frac{1}{3}$.

Unfortunately, although both of the above arguments appear plausible, neither is correct. The reason for this error lies in the model we have assumed. That is, without a doubt $P[E_1 \mid E_2] = \frac{1}{3}$, but this is not exactly what was asked. The question was: What is the probability that TT is correct *given* the student *sees* that the answer to one question is true? That is, we desire to find $P[E_1 \mid \text{student sees one question is true}]$.

Is this not the same as $P[E_1 \mid E_2]$? Not quite, as we see below. Let E_3 be the event that the student sees an answer and E_4 be the event that the answer is true. [Note that we are now varying considerably from the probability space defined by (4) or (5).] Then we desire

$$P[E_1 \mid E_3E_4] = \frac{P[E_1E_3E_4]}{P[E_3E_4]} = \frac{P[E_1]\,P[E_3E_4 \mid E_1]}{P[E_3E_4]} = \frac{P[E_1]\,P[E_3 \mid E_1E_4]}{P[E_3]\,P[E_4 \mid E_3]},$$

since $P[E_3E_4 \mid E_1] = P[E_3 \mid E_1E_4]$. But obviously $P[E_3 \mid E_1E_4] = P[E_3]$, since the possibility of seeing an answer has nothing to do with whether or not it is true or false. Thus

$$P[E_1 \mid E_3E_4] = \frac{P[E_1]}{P[E_4 \mid E_3]}. \tag{6}$$

Now if we assume that the student is equally likely to see a true as a false given that he sees an answer, then $P[E_4 \mid E_3] = \frac{1}{2}$ and

$$P[E_1 \mid E_3E_4] = \frac{\frac{1}{4}}{\frac{1}{2}} = \frac{1}{2}. \tag{7}$$

Hence we have an apparent paradox. For the problem as it has been stated, (7), is the correct answer.

To explain this, let us consider the problem more closely. In obtaining the result (7) we see that the underlying probability space needed for (6) is not the one given by (5), since neither E_3 or E_4 are subsets of that set of outcomes. However, this is not the point we wish to make. The point that merits our attention is the following. Certainly Ω_1 could be considered the sample space of all possible correct answers to the quiz. The problem with defining Ω_1 as the sample space for this experiment is that we do not know what probability should be assigned to the outcomes TT, TF, and FT. If we use conditional probability, then the probabilities assigned to these outcomes will be in the same proportion as they were in Ω. To clarify what has just been said, consider the following heuristic discussion, which leads again to Definition 1.

Returning to Figure 1.1, suppose one desires the probability that E_1 will occur subject to the condition that E_2 is certain to occur. Now E_2 is certain to occur only if we restrict the probability space to the set E_2. Thus we must restrict the set of possible outcomes to E_2. However, if E_2 is to be a probability space, then we must have $P[E_2] = 1$. Thus we must reassign probabilities to the outcomes of E_2 in such a manner that this condition is met. If we now argue that these probabilities should be reassigned such that they are in the same proportion as they were in the original probability space, then we have

$$\rho_i = cp_i,$$

where ρ_i is the new probability corresponding to the originally assigned probability p_i. But

$$1 = \sum_{E_2} \rho_i = c \sum_{E_2} p_i = cP[E_2], \tag{8}$$

and

$$c = \frac{1}{P[E_2]}. \tag{9}$$

In (9) E_2 is of course to be considered as an event in Ω. Thus

$$\rho_i = \frac{p_i}{P[E_2]}.$$

With this reassignment of probabilities E_2 can be considered a probability space and we can calculate the probability of E_1 subject to the constraint E_2 occurs, that is, $P[E_1 \mid E_2]$, by axiom (b), Section 1.4. That is,

$$P[E_1 \mid E_2] = \sum_{E_1E_2} \rho_i = \frac{\sum_{E_1E_2} p_i}{P[E_2]} = \frac{P[E_1E_2]}{P[E_2]}, \tag{10}$$

the latter probabilities referring to the original space. Note however, that (10) is precisely the definition given for

$P[E_1 \mid E_2]$ by (1). Thus in our definition of conditional probability we actually have adopted a mathematical model, and in assuming this model for the conditional probability we have inherently made assumptions on a new probability space. In our present problem the conditional approach will assign equal probabilities to the outcomes TT, TF, and FT, namely $\frac{1}{3}$, since these outcomes all had equal probabilities in the original space $\Omega = \{TT,TF,FT,FF\}$. However, a little reflection easily tells us this is not the model we want. This is seen by noting that the fact the student saw a true is support for the argument that true is more likely than false. Thus the answer TT should not be assigned the same probability as TF. To make this point clearer, suppose the professor spins a sheet 10 times and always leaves the answer to one question exposed. If a student walks by after each spin and notes each time that the exposed answer is true but does not note which answer is exposed, then the possible answers are still TT, TF, and FT. Surely it is clear, however, that TT is much more likely then TF and FT. Hence, it is clear that the fallacy in our previous reasoning is simply that the conditional probability defined on the events of the space $\Omega = \{TT,TF,FT,FF\}$ is simply an incorrect model, and a more comprehensive outcome space should be considered initially.

The engineer should, therefore, keep in mind when using conditional probability that if the given information does not have the same effect on all probabilities of the outcomes within the asserted given event, then the particular conditional probability model he is using is not correct. Thus, in the above, since seeing a true answer offers more credence to TT than TF and FT, we would conclude that the conditional model $P[E_1 \mid E_2]$, $E_2 = \{TT,TF,FT\}$ is an invalid model for that problem.

1.6 STOCHASTIC INDEPENDENCE

In the previous discussion we have hinted at the concept of independence a number of times, for instance, in Example 1, Section 1.4 and again in Example 3, Section 1.5. From those examples we are led to make the following definition.

Definition 1 Two events E_1 and E_2 are said to be independent or stochastically independent if

$$P[E_1 E_2] = P[E_1]\,P[E_2]. \tag{1}$$

Since $P[E_1 E_2] = P[E_1]\,P[E_2 \mid E_1]$ we see that if $P[E_1] > 0$, then E_1 and E_2 are independent if and only if $P[E_2 \mid E_1] = P[E_2]$.

The extension of these ideas to more than two events is as follows.

Definition 2 The n events $E_1, E_2, \ldots, E_n$ are said to be stochastically independent if

$$P[E_1 E_2 \cdots E_k] = P[E_1]\,P[E_2] \cdots P[E_k] \tag{2}$$

for every combination of k events, $1 \leq k \leq n$.

It should be noted that equation (2) is a stronger condition than mutual independence. To illustrate this, consider the following example, which was first given by the Soviet mathematician S. N. Bernstein. Let $\Omega = \{\omega_1, \omega_2, \omega_3, \omega_4\}$ and assume $P[\{\omega_i\}] = \frac{1}{4}$ for each $i = 1, 2, 3, 4$. Now let

$$E_1 = \{\omega_1, \omega_2\}, \qquad E_2 = \{\omega_1, \omega_3\}, \qquad E_3 = \{\omega_1, \omega_4\}.$$

Then

$$P[E_1] = P[E_2] = P[E_3] = \tfrac{1}{2}$$

and

$$E_1 E_2 = E_1 E_3 = E_2 E_3 = \{\omega_1\}.$$

Thus

$$P[E_1 E_2] = P[E_1 E_3] = P[E_2 E_3] = \tfrac{1}{4}.$$

However,

$$P[E_1]\,P[E_2] = P[E_1]\,P[E_3] = P[E_2]\,P[E_3] = \tfrac{1}{4}.$$

Thus

$$P[E_1 E_2] = P[E_1]\,P[E_2],$$
$$P[E_1 E_3] = P[E_1]\,P[E_3],$$

and

$$P[E_2E_3] = P[E_2]P[E_3],$$

so the events are mutually independent. However,

$$E_1E_2E_3 = \{\omega_1\}$$

and hence

$$P[E_1E_2E_3] = \tfrac{1}{4} \neq P[E_1]P[E_2]P[E_3] = \tfrac{1}{8}.$$

Consequently, although E_1, E_2, and E_3 are all jointly independent, they do not satisfy (2) and hence the three events are not independent!

Example 1. Let three fair coins be tossed. Let E_1 = {all heads or all tails}, E_2 = {at least two tails}, and E_3 = {at most two tails}. We note that

$$\begin{aligned} P[E_1] &= \tfrac{2}{8}, \\ P[E_2] &= \tfrac{4}{8}, \\ P[E_3] &= \tfrac{7}{8}, \\ P[E_1E_2] &= \tfrac{1}{8}, \\ P[E_1E_3] &= \tfrac{1}{8}, \end{aligned}$$

and

$$P[E_2E_3] = \tfrac{3}{8},$$

yet

$$\begin{aligned} P[E_1]P[E_2] &= \tfrac{2}{8} \cdot \tfrac{4}{8} = \tfrac{1}{8}, \\ P[E_1]P[E_3] &= \tfrac{2}{8} \cdot \tfrac{7}{8} = \tfrac{14}{64}, \\ P[E_2]P[E_3] &= \tfrac{4}{8} \cdot \tfrac{7}{8} = \tfrac{28}{64}. \end{aligned}$$

It follows that

$$\begin{aligned} P[E_1E_2] &= P[E_1]P[E_2], \\ P[E_1E_3] &\neq P[E_1]P[E_3], \end{aligned}$$

and

$$P[E_2E_3] \neq P[E_2]P[E_3].$$

Hence E_1 and E_2 are independent events and E_1 and E_3 are dependent as are E_2 and E_3.

Example 2. Let E_1 and E_2 be independent events; that is,

$$P[E_1E_2] = P[E_1]P[E_2].$$

II
Probability Distribution and Density Functions

2.1 INTRODUCTION

In this chapter we effectively establish the procedure for determining the probability of an event for more general probability spaces. To do this we will find it necessary to associate with each possible outcome of an experiment a real number. This will be accomplished by the introduction of a special function called a *random variable*. Thus, once we have defined a random variable on the outcomes of the experiment we can consider the outcomes of an experiment as simply real numbers. With this accomplished, it is possible to consider the problem of determining a specific probability function more closely. To this end the cumulative distribution and probability density functions are introduced and the remaining part of the chapter is devoted to these functions and their relation to the random variable.

2.2 RANDOM VARIABLES

Previously we have not restricted the nature of experimental outcomes in any manner. Thus, for example, we could consider four aces and a king as a legitimate outcome in the experiment of drawing five cards from an ordinary deck of cards, or three hits in five passes as an experimental outcome of a bombing run. This broad diversity of the nature of experimental outcomes can be eliminated in a very desirable way by associating each experimental outcome with a real number. In this way we can consider the outcomes of any experiment as real numbers, and therefore make use of existing mathematical structure. Thus, if an energy source emits energy at random during any fixed time period T and if we associate 0 with the event of no energy emitted in time T and 1 with the emission of energy during T, then we can consider the probability space as the set $\{0,1\}$. In actuality it is not necessary to redesignate the sample space, as we may have just inferred. However, it is necessary to associate a real number to each outcome in the original space. This mapping or means of associating is called a random variable. That is, in the above we say that X is a function or random variable defined on Ω such

$$\begin{aligned} X\,(\text{no energy in time } T) &= X\,(\omega_1) = 0, \\ X\,(\text{energy in time } T) &= X\,(\omega_2) = 1, \end{aligned} \tag{1}$$

where the statements in parentheses have been designated by ω_1 and ω_2. Hence, rather than defining a new probability space, we can use the original space by interpreting $P(0)$ and $P(1)$ as

$$P[\omega \in \Omega;\, X(\omega) = 0] \qquad \text{and} \qquad P[\omega \in \Omega;\, X(\omega) = 1], \tag{2}$$

respectively.

These latter statements are read "the probability of an $\omega \in \Omega$ occurring such that $X(\omega) = 0$" and "the probability of an $\omega \in \Omega$ occurring such that $X(\omega) = 1$." The above notation is usually shortened somewhat to the notation $P[X = 0]$ and $P[X = 1]$ rather than equation (2).

2.3 DISCRETE RANDOM VARIABLES

Consider now the following formal definition.

Definition 1 A random variable X is a function defined on Ω whose range is a nonempty set of real numbers and which has the property that $[X \leq x]$ is an event for every real number x.

The notation $[X \leq x]$ is an abbreviated notation for the set of all $\omega \in \Omega$ such that $X(\omega) \leq x$, that is, $\{\omega \in \Omega \mid X(\omega) \leq x\}$. This abbreviation will be used hereafter and should lead to no confusion since it is easily associated with the particular physical problem at hand. That is, when we write $P[X \leq 10]$ or $P[X = 10]$ it will lead to no confusion (in fact, it will probably be less confusing) to simply read this as "the probability $X \leq 10$" and "the probability that $X = 10$" rather than the more technically correct statements "the probability of $\{\omega \in \Omega$ such that $X(\omega) \leq 10\}$" and "the probability of $\{\omega \in \Omega$ such that $X(\omega) = 10\}$."

Capital letters will usually refer to random variables and lower case letters will normally refer to values of random variables, that is, real numbers. We will assume, without proof, that any functions of random variables considered are again random variables.

Definition 2 A random variable X will be said to be discrete if its range is finite or countable, countable meaning it can be put in one-to-one correspondence with the positive integers. By range we mean the set of all values of X.

The above definition simply means that if we consider the set of all possible outcomes of an experiment as a set of real numbers, then when this set of numbers is finite or countable we call the random variable discrete. Hence, in any experiment in which the set of all possible outcomes is finite or countable, we call the associated random variable discrete. Thus, for example, (1), Section 2.2, defines a discrete random variable.

2.4 DISCRETE PROBABILITY DENSITY AND DISTRIBUTION FUNCTIONS

Consider an experiment in which we are only interested in two possible outcomes, for example, firing or nonfiring of a rocket in response to a signal. If we associate 0 with failure (nonfiring) and 1 with success (firing) by means of a random variable X, then the random variable is certainly discrete. Now suppose p is the probability of a success and $q = 1 - p$ is the probability of a failure. Then if

$$f_X(x) = \begin{cases} p^x q^{1-x}, & x = 0,1, \\ 0, & \text{elsewhere,} \end{cases}$$

we see that

$$f_X(0) = P[X = 0] = P[\text{success}],$$
$$f_X(1) = P[X = 1] = P[\text{failure}],$$

and

$$f_X(x) = 0, \qquad \text{if } x \neq 0,1.$$

The function f_X, therefore, is a function which, when evaluated, at any x yields the probability that the random variable will take on that value. Such a function is called a discrete probability density function (pdf).

Definition 1 Let X be a discrete random variable. Then any function f_X such that for every real number x

$$f_X(x) = P[X = x]$$

is called a discrete probability density function. From the above, it is clear that any function such that

$$\text{(a) } f_X(x) \geq 0, \qquad \text{for all } x$$

and

$$\text{(b) } \sum_{n=0}^{\infty} f_X(x_n) = 1, \qquad \text{where } f_X(x) = 0 \text{ if } x \neq x_n,$$

can serve as a discrete probability density function. At present the subscript X may seem superfluous to the reader, but its value will become clear when we consider multiple density functions.

In most cases the discrete random variables we will consider will take only integer values, that is, $0, 1, 2, \ldots.$ In this case, conditions (a) and (b) become $f_X(n) \geq 0$ and $\sum_{n=0}^{\infty} f(n) = 1$. Also we have

$$P[X \leq k] = \sum_{n=0}^{k} f_X(n). \tag{1}$$

Consider the experiment of tossing four coins. If x is the number of heads obtained and

$$f_X(x) = \begin{cases} \binom{4}{x}\left(\frac{1}{2}\right)^4 = \frac{4!}{x!(4-x)!}\left(\frac{1}{2}\right)^4, & \text{for } x = 0, 1, 2, 3, 4, \\ 0, & \text{elsewhere,} \end{cases}$$

we see that

$$f_X(0) = \left(\frac{1}{2}\right)^4 = P[X = 0],$$

$$f_X(1) = 4\left(\frac{1}{2}\right)^4 = P[X = 1],$$

$$f_X(2) = 6\left(\frac{1}{2}\right)^4 = P[X = 2],$$

etc.

Clearly, $f_X(x) \geq 0$ and

$$\sum_{n=0}^{\infty} f_X(n) = \sum_{n=0}^{4} f_X(n) = \sum_{n=0}^{4} P[X = n]$$

$$= P\sum_{n=0}^{4} [X = n] = P[\Omega] = 1.$$

Thus $f_X(x)$ is a discrete probability density function.

Example 1. The function $f_X(x)$ in the above illustration is a special case of the function defined by ($0 \leq p \leq 1$),

$$f_X(x) = \begin{cases} \binom{n}{x} p^x q^{n-x}, & x = 0, 1, \ldots, n, p + q = 1, \\ 0, & \text{elsewhere.} \end{cases}$$

This particular function, as we shall see in Example 5, is an important probability model for a large class of problems. In this example we will show $f_X(x)$ is a pdf. It is clear that $f_X(x) \geq 0$, and hence we need only show that

$$\sum_{x=0}^{n} f_X(x) = 1.$$

But

$$\sum_{x=0}^{n} \binom{n}{x} p^x q^{n-x} = (p + q)^n$$

by the binomial theorem, and since $p + q = 1$, we have the desired result.

Example 2. It has been verified experimentally that the number of failures in a large system during a fixed time T can often be described by the probability model given by the pdf

$$f_X(x) = \begin{cases} e^{-\mu}\left(\dfrac{\mu^x}{x!}\right), & x = 0, 1, 2, \ldots, \\ 0, & \text{elsewhere.} \end{cases}$$

Find the probability of
(a) exactly three failures,
(b) at most three failures, and
(c) at least four failures in time T.
Assume $\mu = 0.1$.

Solution:

(a) Since f_X is a discrete density function,

$$P[X = 3] = f_X(3) = \frac{(.1)^3 e^{-(.1)}}{3!} = \frac{(.001)e^{-(.1)}}{6}.$$

(b) Since at most, three failures is composed of the events 0, 1, 2, or 3 failures, and since these events are mutually exclusive, we have

$$\begin{aligned} P[X \leq 3] &= P[X = 0] + P[X = 1] + P[X = 2] + P[X = 3] \\ &= \sum_{n=0}^{3} \frac{e^{-(.1)}(.1)^n}{n!}. \end{aligned}$$

(c) We must have at most three failures or at least four failures in T. That is, at most three failures occur if we have 0, 1, 2, or 3 failures, and at least four failures occur if we have 4, 5, 6, 7, ... failures in T. Since this is the entire range of the random variable, and since these are mutually exclusive, we have

$$P[X \leq 3] + P[X \geq 4] = 1.$$

Thus

$$P[X \geq 4] = 1 - P[X \leq 3]$$
$$= 1 - \sum_{n=0}^{3} \frac{e^{-(.1)}(.1)^n}{n!}.$$

From the above one sees that when the probability density function is known, the question of how one assigns numbers to the probability function is solved. It is also seen, however, that the probability density function will normally be determined through considerations of a priori probability (Example 1) or relative frequency (Example 2).

Example 3. Determine c such that if

$$f_X(x) = \begin{cases} cx, & x = 1,2, \\ 0, & \text{elsewhere,} \end{cases}$$

then f_X is a probability density function.

Solution: We must have $f_X(x) \geq 0$ and

$$\sum_{x=1}^{2} f_X(x) = c \sum_{x=1}^{2} x = 3c = 1.$$

Therefore, if $c = \frac{1}{3}$, we see that f_X is a density function.

Example 4. Can the function defined by

$$f_X(x) = \begin{cases} c + x, & x = 0,1,2, \\ 0, & \text{elsewhere} \end{cases}$$

be a probability density function?

Solution: We must have

$$\sum_{x=0}^{2} f_X(x) = 3c + 3 = 1$$

or

$$c = -\tfrac{2}{3}.$$

But then $f_X(0) < 0$, and hence f_X cannot be a probability density function. Note that the function $f_X(x) = c + x$ for $x = 1, 2$, and $f_X(x) = 0$ elsewhere can serve as a pdf.

In several of the above examples we have considered the function $P[X \leq x]$. In general, this function plays a very important role in the field of probability. It is called the *distribution function* or cumulative distribution function. The latter name is usually used to distinguish it from the probability density function, which is often also referred to as a distribution function.

Definition 2 Let P be a probability function and X a random variable. Then the function F_X defined by

$$F_X(x) = P[X \leq x] \tag{2}$$

is called the probability distribution function or cumulative probability distribution function. The following properties of the distribution function are obvious:

(a) $F_X(\infty) = 1$.
(b) $F_X(-\infty) = 0$.
(c) $F_X(x) \geqq 0$.
(d) If $x_1 < x_2$, then $F_X(x_1) \leq F_X(x_2)$.
(e) If X is a discrete random variable with range $\{x_n\}$ such that $x_0 \leq x_1 \leq x_2 \leq \cdots$, then

$$F_X(x) = \sum_{n=0}^{[x]} f(x_n),$$

where $[x]$ is the largest integer such that x_n does not exceed x.

Example 5. Suppose an electron gun fires n electrons in succession at a target. If the probability that an electron will hit its target is p, find the probability that exactly x of the electrons will hit the target. Assume that hitting the target is an independent phenomenon.

Solution: Consider first the probability that the first x electrons fired hit the target and the remaining $n - x$ fail to hit it. In this case, we see

$$P[\text{first } x \text{ hit and next } n - x \text{ miss}] = p^x q^{n-x}, \qquad q = 1 - p.$$

However, we have considered only one way in which x out of n electrons can hit the target. Clearly, there are a number of other ways, each equally likely. Moreover, all these orders are certainly mutually exclusive. But the total number of ways of arranging two distinct groups of x indistinguishable objects with $n - x$ indistinguishable objects is $\binom{n}{x}$. Therefore,

$$f_X(x) = \begin{cases} \binom{n}{x} p^x q^{1-x}, & x = 0, 1, \ldots, n, \\ 0, & \text{elsewhere}, \end{cases} \tag{3}$$

and

$$P[\text{exactly } k \text{ hits}] = f_X(k).$$

Moreover,

$$P[X \leq x] = F_X(x)$$

and

$$P[x_1 < X < x_2] = F_X(x_2 - 1) - F_X(x_1).$$

Thus the probabilities of less than three hits and greater than three but less than eight hits ($n \leq 7$) are $F_X(3)$ and $F_X(7) - F_X(4)$, where

$$F_X(3) = \sum_{x=0}^{3} \binom{n}{x} p^x q^{n-x}$$

and

$$F_X(7) - F_X(3) = \sum_{x=4}^{x=7} \binom{n}{x} p^x q^{n-x}.$$

The above example suggests the following easily proved and useful theorem.

Theorem 1 Let X be a random variable whose range

is contained in the set $\{0, 1, 2, \ldots\}$. Then

(a) $P[n_1 < X < n_2] = F_X(n_2 - 1) - F_X(n_1)$
(b) $P[n_1 \leq X < n_2] = F_X(n_2 - 1) - F_X(n_1 - 1)$
(c) $P[n_1 < X \leq n_2] = F_X(n_2) - F_X(n_1)$
(d) $P[n_1 \leq X \leq n_2] = F_X(n_2) - F_X(n_1 - 1)$

2.5 MULTIVARIATE DISCRETE DENSITY AND DISTRIBUTIONS

In order to develop satisfactory probability models for joint events it is necessary to extend the above analysis somewhat.

Definition 1 Let $X_1, X_2, \ldots, X_n$ be discrete random variables. A multivariate discrete density function is any function $f_{X_1 \cdots X_n}$ such that

$$f_{X_1 \cdots X_n}(x_1, x_2, \ldots, x_n) = P[X_1 = x_1, X_2 = x_2, \ldots, X_n = x_n]$$

for every n-tuple of real numbers $(x_1, x_2, \ldots, x_n)$. (Comment: An n-component vector is often called an n-tuple.)

In Definition 1 we see that a multivariate density is simply a function that gives the probability of a compound event. For example, suppose x = number of aces, y = number of kings in a five-card poker hand. Let

$$f_{XY}(x, y) = \frac{\binom{4}{x}\binom{4}{y}\binom{44}{5 - x - y}}{\binom{52}{5}},$$

where we define $\binom{n}{k} = 0$, if $k < 0$ or $k > n$. Then

$$f_{XY}(0,0) = \frac{\binom{44}{5}}{\binom{52}{5}} = P[X = 0, Y = 0]$$

and

$$f_{XY}(1,3) = \frac{\binom{4}{1}\binom{4}{3}\binom{44}{1}}{\binom{44}{5}} = P[X = 1, Y = 3].$$

It is clear, in general, that

$$f_{XY}(x,y) = P[X = x, Y = y],$$

and hence f_{XY} is a joint or multivariate probability density function. Note that $f(0,5) = 0$, which is in agreement with the fact that zero aces and five kings is an impossible event.

Definition 2 If $X_1, \ldots, X_n$ are random variables, then we define the conditional density $f_{X_1 \cdots X_k \mid X_{k+1} \cdots X_n}$ by

$$f_{X_1 \cdots X_k \mid X_{k+1} \cdots X_n}(x_1, x_2, \ldots, x_k \mid x_{k+1} \cdots x_n)$$
$$= P[X_1 = x_1, \ldots, X_k = x_k \mid x_{k+1} \cdots x_n]$$

for every n-tuple of real numbers $(x_1, \ldots, x_n)$.

From the above definition and Theorem 1, Section 1.5, we see that

$$\begin{aligned} f_{X_1X_2}(x_1,x_2) &= f_{X_1}(x_1)f_{X_2 \mid X_1}(x_2 \mid x_1) \\ &= f_{X_2}(x_2)f_{X_1 \mid X_2}(x_1 \mid x_2), \qquad (1) \\ f_{X_1X_2X_3}(x_1,x_2,x_3) &= f_{X_1}(x_1)f_{X_2 \mid X_1}(x_2 \mid x_1)f_{X_3 \mid X_1X_2}(x_3 \mid x_1x_2), \end{aligned}$$

etc.

Note that if R is the range of say, X_2, then

$$\sum_R f_{X_1X_2}(x_1,x_2) = f_{X_1}(x_1)\sum_R f_{X_2 \mid X_1}(x_2 \mid x_1)$$
$$= f_{X_1}(x_1)(1) = f_{X_1}(x_1).$$

The above will be clear by considering once more the joint density

$$f_{XY}(x,y) = \frac{\binom{4}{x}\binom{4}{y}\binom{44}{5 - x - y}}{\binom{52}{5}}.$$

Suppose $x = 3$. Then

$$f_{XY}(3, y) = \frac{4\binom{4}{y}\binom{44}{2-y}}{\binom{52}{5}}.$$

Now

$$\begin{aligned}\sum_{y=0}^{4} f_{XY}(3, y) &= P[3 \text{ aces and } 0 \text{ kings}] + P[3 \text{ aces and } 1 \text{ king}] \\ &\quad + P[3 \text{ aces and } 2 \text{ kings}] \\ &= P[3 \text{ aces}]\left\{\sum_{i=0}^{2} P[i \text{ kings} \mid 3 \text{ aces}]\right\}.\end{aligned}$$

But

$$\sum_{i=0}^{2} P[i \text{ kings} \mid 3 \text{ aces}] = 1,$$

since if it is given that we have three aces, then we must have 0, 1, or 2 kings. It follows then that

$$f_X(3) = \sum_{y=0}^{4} f_{XY}(3, y) = \sum_{y=0}^{2} f_{XY}(3, y),$$

or, for any x,

$$f_X(x) = \sum_{y=0}^{4} f_{XY}(x, y) = \sum_{y=0}^{5-x} f_{XY}(x, y).$$

This property of joint densities is quite useful and we now state it as a theorem.

Theorem 1 If $X_1 \cdots X_n$ are discrete random variables with the pdf $f_{X_1 \cdots X_n}$, then $X_1, \ldots, X_k$, $k < n$, are discrete random variables with the pdf $f_{X_1 \cdots X_k}$ given by

$$f_{X_1 \cdots X_k}(x_1, \ldots, x_k) = \sum_{R} f_{X_1 \cdots X_n}(x_1, x_2, \ldots, x_n), \tag{2}$$

where R is the range of the random variables $X_{k+1}, \ldots, X_n$.

The density function on the left side of equation (2) is often called the *marginal probability density function.*

Example 1. Find the probability density function for the number of aces in a five-card poker hand.

Solution: Since the number of ways one can draw x aces is $\binom{4}{x}$ and the remaining cards can be drawn is $\binom{48}{5-x}$, we see that

$$f_X(x) = \begin{cases} \dfrac{\binom{4}{x}\binom{48}{5-x}}{\binom{52}{5}}, & 0 \le x \le 4, \\ 0, & \text{elsewhere.} \end{cases}$$

An alternative approach is by using the joint density function for aces and kings previously discussed. That is,

$$f_X(x) = \sum_{y=0}^{5-x} f_{XY}(x, y)$$

$$= \frac{\binom{4}{x}}{\binom{52}{5}} \sum_{y=0}^{5-x} \binom{4}{y}\binom{44}{5-x-y}.$$

It is well known that

$$\sum_{i=0}^{c} \binom{a}{i}\binom{b}{c-i} = \binom{a+b}{c}.$$

Hence

$$\sum_{y=0}^{5-x} \binom{4}{y}\binom{44}{5-x-y} = \binom{48}{5-x},$$

and again

$$f_X(x) = \frac{\binom{4}{x}\binom{48}{5-x}}{\binom{52}{5}}.$$

Example 2. In the above example, suppose it is given that the hand contains two aces. What is the probability it also contains three kings?

Solution: If x is the number of aces and y is the number of kings the hand contains, then

$$f_{Y|X}(y \mid x) = \frac{f(x,y)}{f(x)} = \frac{\binom{4}{x}\binom{4}{y}\binom{44}{5-x-y}}{\binom{52}{5}} \cdot \frac{\binom{52}{5}}{\binom{4}{x}\binom{48}{5-x}}$$

$$= \frac{\binom{4}{y}\binom{44}{5-x-y}}{\binom{48}{5-x}}.$$

Thus

$$f_{Y|X}(3 \mid 2) = \frac{\binom{4}{3}\binom{44}{0}}{\binom{48}{3}} = \frac{1}{2 \cdot 47 \cdot 46},$$

which is the probability of three kings given two aces.

Note from above that, theoretically, if the joint density function is known, questions of probability associated with the experiment are easily answered.

2.6 INDEPENDENCE OF RANDOM VARIABLES

We have previously discussed the concept of independent events. We will now extend this concept to discrete random variables in the obvious manner. That is, since discrete densities are no more than probabilities, we say that the n discrete random variables $X_1, X_2, \ldots, X_n$ are independent if

$$f_{X_1 \cdots X_n}(x_1, x_2, \ldots, x_n) = f_{X_1}(x_1) f_{X_2}(x_2) \cdots f_{X_n}(x_n) \tag{1}$$

for every n-tuple of real numbers $(x_1, x_2, \ldots, x_n)$.

It is easy to see that if two random variables X_1 and X_2 are independent, then

$$f_{X_1 \mid X_2}(x_1 \mid x_2) = f_{X_1}(x_1)$$

and

$$f_{X_2 \mid X_1}(x_2 \mid x_1) = f_{X_2}(x_2).$$

In most physical cases we do not use equation (1) to determine the physical independence of experimental outcomes but rather the converse. That is, if experience dictates that the outcomes of two experiments are independent, then we adopt equation (1) as the appropriate probability model. This, of course, is a sound procedure, since the observation of independent phenomena motivated our original definition of independence.

2.7 CONTINUOUS RANDOM VARIABLES, DENSITIES, AND DISTRIBUTIONS

We have given a general definition of a distribution function in Definition 2, Section 2.4. At that time we noted that if X was a discrete random variable such that $x_0 \leq x_1, \ldots$, then

$$F_X(x) = \sum_{n=0}^{[x]} f_X(x_n),$$

and we called $F_X(x)$ a discrete distribution function. We will now consider an extension of that idea which is useful when the outcomes of an experiment are not countable.

Definition 1 If there exists a function f_X such that

$$F_X(x) = \int_{-\infty}^{x} f_X(t)\,dt \tag{1}$$

for every real number x, then we say that F_X is a *continuous distribution function.* The function f_X is then referred to as a continuous probability density function for X. From (1) we see that if f_X is a continuous density function, then $P[X \leq x_0] =$

$F_X(x_0)$, which equals the area under the curve $f_X(x)$ to the left of x_0.

Definition 2 If the random variable X has a continuous probability density function, then we call X a *continuous random variable.*

From an engineer's point of view, that is, in the physical world, one can simply consider a continuous random variable as one which can take on an uncountable number of values or all the values in an interval.

The following are examples of values for which we would normally describe the associated random variable as continuous:

1. The height of a man selected at random.
2. The length of a bolt selected at random from a box of mixed bolts.
3. The lengths of "4-inch" bolts selected at random from a box of "4-inch" bolts.

All the above would be described as the value of a continuous random variable, since the set of possible values in each case is uncountable. That is, if one selects a bolt from a box of "4-inch" bolts, the bolt may be anywhere from 4.01 inches to 3.99 inches. But this means the length of the bolt could be any number in the interval (3.99, 4.01), and hence the associated random variable is usually called continuous.

Although in the mathematical theory of probability continuous and discrete random variables form only a small subclass of the set of all random variables, we will have no need in this book for any other types of random variables. In fact, the practicing engineer will encounter no difficulty if he considers all random variables which are not discrete as continuous random variables. This will probably avoid any confusion and is possibly the manner in which we should have introduced continuous random variables. However, to avoid making bold false statements, even in the name of expediency, we introduced this concept in the manner above which leads to no false statements and, it is hoped, no significant confusion.

By Definition 1 we see that any function f_X which satisfies the conditions

$$\text{(a) } f_X(x) \geq 0,$$

$$\text{(b)} \quad \int_{-\infty}^{\infty} f_X(x)\,dx = 1$$

can serve as a continuous density function.

These ideas are extended to the multivariate case by the following definition.

Definition 3 If there exists a function $f_{X_1X_2\cdots X_n}$ such that

$$F_{X_1X_2\cdots X_n}(x_1, x_2, \ldots, x_n) = \int_{-\infty}^{x_1} \cdots \int_{-\infty}^{x_n} f_{X_1\cdots X_n}(t_1, \ldots, t_n)\,dt_n \cdots dt_1 \qquad (2)$$

for every n-tuple of real numbers $(x_1, \ldots, x_n)$, then $F_{X_1\cdots X_n}$ is called a continuous multivariate distribution function and $f_{X_1\cdots X_n}$ is called a continuous multivariate density function. Further, when (2) holds, the random variables $X_1, \ldots, X_n$ will be referred to as continuous random variables.

Definition 4 If X and Y are continuous random variables, we define the conditional density function of X given Y, that is, $f_{X|Y'}$ by

$$f_{X|Y}(x \mid y) = \frac{f(x, y)}{f(y)}, \qquad (3)$$

provided $f(y) > 0$. The function

$$F_{X|Y}(x \mid y) = \int_{-\infty}^{x} f_{X|Y}(t \mid y)\,dt$$

is then called a *conditional distribution function.*

Definition 4 is not the most general or even the best one possible to give for the conditional density function. It is, however, satisfactory for our purposes and has the advantage of simplicity. For a more complete discussion see Tucker (Ref. 21).

In the above it should be noted that we have extended the ideas of discrete sample spaces and random variables to a more general sample space in which the outcomes are no longer finite or countable and to continuous random variables. This was done in a most natural way and possibly the transition was so intuitively obvious that the reader may have overlooked the fact that there are some resulting distinctions between continuous and discrete density function. For example, if X is a discrete random variable and f_X is the associated density function, then $f_X(x)$ is a probability for every real number x,

$$f_X(x) = P[X = x].$$

This is not true for continuous density functions. In fact, if X is a continuous random variable, then

$$P[X = x] = 0$$

for every real number x. This would follow from (1) by noting

$$P[X \leq x] = P[X < x],$$

since the value of the integral is not changed by deleting one point. But

$$P[X \leq x] = P[X < x] + P[X = x],$$

and hence $P[X = x] = 0$. A little reflection shows that this is a reasonable result from a physical point of view as well. That is, it is reasonable that the probability that a man, selected at random, will be exactly 6 feet tall is zero. This is true since upon the use of a sufficiently fine measuring device, we would find that no man is exactly 6 feet tall—maybe somewhere between 6.01 and 5.99 feet or even 6.000001 feet and 5.999999 feet, but not exactly 6 feet. Of course, $P[5.999999 < X < 6.000001] \neq 0$, necessarily but is given by $F(6.000001) - F(5.999999)$. Thus we see that our extension has resulted in a meaningful physical model.

We now list the corresponding theorem to Theorem 1, Section 2.5, for continuous density functions.

Theorem 1 If $X_1 \cdots X_n$ are continuous random variables with pdf $f_{X_1 \cdots X_n}$, then, if $k < n$, $X_1, \ldots, X_k$ are continuous random variables with pdf $f_{X_1 \cdots X_k}$ given by

$$f_{X_1 \cdots X_k}(x_1, \ldots, x_k)$$

$$= \int_{-\infty}^{\infty} \cdots \int_{-\infty}^{\infty} f_{X_1 \cdots X_n}(t_1, \ldots, t_n)\, dt_n \cdots dt_{k+1}.$$

Discrete and continuous random variables, densities, and distributions are summarized in Table 2.1. In Table 2.1 statements in regard to the random variable X in the left column should be interpreted as the left side of an equality; the two columns on the right form the right side of the equality accordingly as X is discrete or continuous.

Before closing this section we should mention the important notion of stochastic independence of random variables. We have defined this concept for discrete random variables. However, the same physical notions which led us to (1), Section 2.6, would lead us to a similar definition in the case of continuous random variables, which we now give.

Definition 5 Let $f_{X_1}, f_{X_2}, \ldots, f_{X_n}$ be continuous probability density functions for the continuous random variables $X_1 X_2, \ldots, X_n$. Then $X_1, \ldots, X_n$ are said to be independent if and only if the multivariate (or joint) density $f_{X_1 X_2 \cdots X_n}$ is given by

$$f_{X_1 X_2 \cdots X_n}(x_1, \ldots, x_n) = f_{X_1}(x_1)\, f_{X_2}(x_2) \cdots f_{X_n}(x_n) \tag{4}$$

for every n-tuple of real numbers $(x_1, \ldots, x_n)$.

A concept closely associated with independence and one we shall use extensively is that of a random sample. A random sample is simply a set of independent random variables from the same sample space. Thus, in light of Definition 5 and (1), Section 2.6, we have the following definition.

Definition 6 The set of random variables $\{X_1, \ldots, X_n\}$ is called a random sample if and only if they are independent and have the same probability density function, that is, if and only if

$$f_{X_1 X_2 \cdots X_n}(x_1, \ldots, x_n) = f_X(x_1)\, f_X(x_2) \cdots f_X(x_n). \tag{5}$$

We conclude this section with the following examples.

Table 2.1

Statement =	Discrete Random Variables	Continuous Random Variables
$P[X \leq x]$, x a real number	$F_X(x) = \sum_{n=0}^{[x]} f_x(x_n)$*	$F_X(x) = \int_{-\infty}^{x} f_x(t)\,dt$
$P[n_1 \leq X \leq n_2]$	$F_X(n_2) - F_X(n_1 - 1)$	$F_X(n_2) - F_X(n_1)$
$P[n_1 < X < n_2]$	$F_X(n_2 - 1) - F(n_1)$	$F_X(n_2) - F_X(n_1)$
$P[X = x_1]$, x_1 any real number	$f_X(x_1)$	0
Conditional density function of X given Y	$f_{X\mid Y}(x \mid y) = \dfrac{f_{XY}(x, y)}{f_Y(y)}$, $f_Y(y) > 0$	$f_{X\mid Y}(x \mid y) = \dfrac{f_{XY}(x, y)}{f(y)}$, $f_Y(y) > 0$
$P[X = x \mid Y = y]$	$f_{X\mid Y}(x \mid y) = \dfrac{f_{XY}(x, y)}{f_Y(y)}$	0
$P[X \leq x \mid Y = y]$	$\sum_{n=0}^{[x]} f_{X\mid Y}(x_n \mid Y)$*	$\int_{-\infty}^{x} f_{X\mid Y}(t \mid y)\,dt$
Marginal density of X, f_X, associated with the joint density f_{XY}	$f_X(x) = \sum_Y f_{XY}(x, y)$	$f_X(x) = \int_{-\infty}^{\infty} f_{XY}(x, y)\,dy$
$P[X \leq -\infty]$	$F_X(-\infty) = 0$	$F_X(-\infty) = 0$
$P[X \leq \infty]$	$F_X(\infty) = 1$	$F_X(\infty) = 1$
$x_1 \leq x_2$	$F_X(x_1) \leq F_X(x_2)$	$F_X(x_1) \leq F_X(x_2)$
Range of values of X	Finite or countable	Uncountable or all values in an interval

*Here and, in fact, henceforth, we assume that $x_0 \leq x_1 \leq x_2, \ldots$ and $[x]$ is the largest integer such that $x_n \leq x$.

Example 1. Let X and Y be two random variables whose associated joint probability density function is

$$f_{XY}(x, y) = \begin{cases} 1, & 0 \leq x \leq 1, 0 \leq y \leq 1, \\ 0, & \text{elsewhere.} \end{cases}$$

(a) Evaluate $P[X > Y]$.

(b) Find the marginal probability density functions for Y and X.
(c) Find the conditional probability density of X given Y.
(d) Comment on the dependence or independence of X and Y.

Solution: (a) Consider

$$P[X > Y] = \iint_R f_{XY}(x, y)\, dx\, dy$$

where $R = \{(x, y); x > y\}$. That is,

$$\begin{aligned} P[X > Y] &= \int_0^1 \int_y^1 1\, dx\, dy \\ &= \int_0^1 (1 - y)\, dy \\ &= y - \tfrac{1}{2} y^2 \Big|_0^1 \\ &= 1 - \tfrac{1}{2} \\ &= \tfrac{1}{2}. \end{aligned}$$

(b) The marginal probability density function of Y is, by definition,

$$\begin{aligned} f_Y(y) &= \int_{-\infty}^{\infty} f_{XY}(x, y)\, dx \\ &= \int_0^1 1\, dx \\ &= 1, \qquad 0 < y < 1. \end{aligned}$$

By symmetry $f_X(x) = 1, 0 < x < 1$.

(c) The conditional probability density function of X given Y is, by definition,

$$\begin{aligned} f_{X|Y}(x \mid y) &= \frac{f_{XY}(x, y)}{f_Y(y)} \\ &= \frac{1}{1} \\ &= 1, \qquad 0 \le x \le 1. \end{aligned}$$

(d) By observing that

$$f_{XY}(x, y) = f_X(x)\, f_Y(y),$$

one concludes that X and Y are independent random variables.

Example 2. Let

$$f_{XY}(x, y) = \begin{cases} 2, & 0 < x < 1, 0 < y < x, \\ 0. \end{cases}$$

Answer the same questions listed in Example 1.

Solution: (a) $P[X > Y] = 1$, since y is always less than x by the statement of the problem.

(b) By definition, the marginal pdf of X is

$$\begin{aligned} f_X(x) &= \int_{-\infty}^{\infty} f_{XY}(x, y)\, dy \\ &= \int_0^x 2\, dy \\ &= 2x, \qquad 0 < x < 1. \end{aligned}$$

Also,

$$\begin{aligned} f_Y(y) &= \int_{-\infty}^{\infty} f_{XY}(x, y)\, dx \\ &= \int_y^1 2\, dx \\ &= 2(1 - y), \qquad 0 < y < 1. \end{aligned}$$

(c) The conditional probability density function of X given Y is

$$\begin{aligned} f_{X \mid Y}(x \mid y) &= \frac{f_{XY}(x, y)}{f_Y(y)} \\ &= \frac{2}{2(1 - y)} \\ &= \frac{1}{1 - y}, \qquad 0 < y < x < 1. \end{aligned}$$

(d) We observe that

$$\frac{1}{1-y} \neq (2x)[2(1-y)]$$

or

$$f_{XY}(x,y) \neq f_X(x) f_Y(y),$$

and hence X and Y are indeed not independent random variables.

Example 3. Let X and Y be two random variables whose joint probability density function is

$$f_{XY}(x,y) = (2\pi)^{-1} \exp[-\tfrac{1}{2}(x^2 + y^2)],$$

$-\infty < x < \infty$ and $-\infty < y < \infty$.

(a) Find the marginal probability density function.

(b) Find the conditional probability density function of Y given X.

(c) Comment on the dependence or independency of X and Y.

Solution: (a) We note that $f_{XY}(x,y)$ can be written in the form

$$f_{XY}(x,y) = [(2\pi)^{-1/2} \exp(-\tfrac{1}{2}x^2)]\ [(2\pi)^{-1/2} \exp(-\tfrac{1}{2}y^2)].$$

Also one can show that

$$1 = \int_{-\infty}^{\infty} (2\pi)^{-1/2} \exp(-\tfrac{1}{2}t^2)\, dt.$$

Hence

$$f_X(x) = (2\pi)^{-1/2} \exp(-\tfrac{1}{2}x^2) \int_{-\infty}^{\infty} (2\pi)^{-1/2} \exp(-\tfrac{1}{2}y^2)\, dy$$

$$= (2\pi)^{-1/2} \exp(-\tfrac{1}{2}x^2).$$

Also

$$f_Y(y) = (2\pi)^{-1/2} \exp(-\tfrac{1}{2}y^2).$$

(b) By definition

$$f_{Y|X}(y \mid x) = \frac{f_{XY}(x,y)}{f_X(x)}$$
$$= (2\pi)^{-1/2} \exp\left(-\tfrac{1}{2} y^2\right).$$

(c) We note that

$$f_{Y|X}(y) = f_Y(y),$$

and hence X and Y are stochastically independent.

Example 4. A bolt machine makes bolts with diameter X; a second machine makes nuts for the bolts with inside diameter Y. Suppose the density of X and Y is

$$f_{XY}(x,y) = \begin{cases} 2500, & .49 < x < .51, .51 < y < .53 \\ 0, & \text{elsewhere.} \end{cases}$$

A nut fits a bolt satisfactorily if its diameter exceeds that of the bolt by at least .002 but not more than .03. What is the probability that a bolt and a nut chosen at random will fit?

Solution: (a) The probability p that a bolt and nut chosen at random will fit can be stated mathematically as

$$p = P[.002 \le Y - X \le .03]$$
$$= \iint_R f_{XY}(x,y)\, dx\, dy,$$

where $R = \{(x,y);\ .002 \le y - x \le .03,\ .49 \le x \le .51, \text{ and } .51 \le y \le .53\}$. Hence

$$p = \int_{.49}^{.50} \int_{.51}^{X+.03} 2500\, dy\, dx$$
$$+ \left(\int_{.50}^{.51} \int_{.51}^{.53} 2500\, dy\, dx - \int_{.508}^{.51} \int_{.51}^{x+.002} 2500\, dy\, dx \right).$$

The complicated limits result from deducing the boundaries of the region R (x,y). The value of p can be obtained by direct integration.

Example 5. Let the North–South miss distance from a target be X, and the East–West miss distance be Y for a specified artillery piece. Let the joint probability density function of X and Y be

$$f_{XY}(x,y) = (2\pi)^{-1}\exp\left[-\tfrac{1}{2}(x^2 + y^2)\right], \qquad -\infty < x < \infty, -\infty < y < \infty.$$

The target is destroyed if $X^2 + Y^2 \leq 1$, that is, if a round from the artillery piece is within one unit of the target. Find the number of independently fired rounds necessary so that one can say with probability .99 that the target is destroyed.

Solution: The probability that the target is destroyed when only one round is fired is

$$\begin{aligned} p &= P[X^2 + Y^2 \leq 1] \\ &= \iint_R f_{XY}(x,y)\,dx\,dy, \end{aligned}$$

where

$$R = \{(x,y); x^2 + y^2 \leq 1\}.$$

Thus,

$$p = \iint_{x^2+y^2\leq 1} f_{XY}(x,y)\,dx\,dy = \int_0^{2\pi}\int_0^1 e^{-(1/2)\gamma^2}\gamma\,d\gamma\,d\theta \approx .393.$$

Now the probability q of a miss is given by

$$\begin{aligned} q &= 1 - p \\ &= 1 - .393 \\ &= .607. \end{aligned}$$

Hence the probability of at least one hit in n trials is given by $1 - q^n$, or

$$P[\text{at least one hit}] = 1 - .607^n.$$

But we require that

$$P[\text{at least one hit}] = .99$$

or

$$\begin{aligned} 1 - .607^n &= .99 \\ .01 &= .607^n \\ \log .01 &= n \log .607 \end{aligned}$$

or, finally,

$$n = \frac{\log .01}{\log .607},$$

and we select the smallest integer greater than the value computed for n as our solution.

2.8 FUNCTIONS OF RANDOM VARIABLES

Suppose we are interested in determining the probability that a missile will hit within a certain circular region, r. Moreover, suppose that, owing to properties of the guidance system, we can reasonably assume as known the joint probability density function for X, the horizontal target error, and Y, the vertical target error. Then our problem is the following. Find the $P[X^2 + Y^2 < r^2]$ if f_{XY} is known. If $Z = X^2 + Y^2$, the problem is to find the distribution or density function of Z. Hence we see that in practice one is quite often confronted with the question: If $f_{X_1 \ldots X_n}$ is known, what is the pdf of $Z = g(X_1, \ldots, X_n)$?" We shall first consider this problem in the case of a single random variable and then extend it.

Theorem 1 Let X be a continuous random variable such that $f_X(x) = 0$ if x is not in the interval (a, b). Let $y = g(x)$ be a one-to-one mapping[1] of x from (a, b) to (c, d) such that dy/dx exists and is nonzero when y is an element of (c, d). Then

---

[1]A mapping $y = g(x)$ is said to be one to one if for each x there corresponds a unique y and vice versa.

$$f_Y(y) = \begin{cases} f_X[g^{-1}(y)] \left| \dfrac{dx}{dy} \right|, & \text{if } y \text{ is an element of } (c, d), \\ 0, & \text{elsewhere,} \end{cases}$$

where g^{-1} denotes the inverse function of g.

Example 1. Suppose

$$f_X(x) = \begin{cases} e^{-x}, & 0 \le x < \infty, \\ 0, & -\infty < x < 0. \end{cases}$$

Find the pdf of the random variable $Y = 3X + 2$. Applying Theorem 1 with $g(x) = 3X + 2$ gives

$$f_Y(y) = \begin{cases} \frac{1}{3} \exp\left[-\left(\dfrac{y-2}{3}\right)\right], & 2 \le y < \infty, \\ 0, & \text{elsewhere.} \end{cases}$$

Example 2. Suppose again $f_X(x) = e^{-x}$ for nonnegative x and is zero elsewhere. Find the pdf of $Y = X^2$. Again the mapping is one to one for all x where $f_X(x)$ is different from zero. Hence

$$f_Y(y) = \begin{cases} \dfrac{1}{2\sqrt{y}} e^{-\sqrt{y}}, & 0 \le y < \infty, \\ 0, & -\infty < y < 0. \end{cases}$$

Example 3. Let $f_X(x) = e^{-(1/2)x^2}$, $-\infty < x < \infty$. Find the pdf of Y if $Y = X^2$. In this case the mapping is not one to one and Theorem 1 does not apply. However, we can solve the problem in the following way. If $y \le 0$, then $F_Y(y) = 0$, but if $0 < y < \infty$, then

$$\begin{aligned} F_Y(y) = P[Y \le y] &= P[X^2 \le y] = P[-\sqrt{y} \le X \le \sqrt{y}] \\ &= F_X(\sqrt{y}) - F_X(-\sqrt{y}) \\ &= \int_{-\sqrt{y}}^{\sqrt{y}} e^{-(1/2)x^2}\, dx. \end{aligned}$$

But $F'_Y(y) = f_Y(y)$, and hence

$$f_Y(y) = \frac{d}{dy}\left\{\int_{-\sqrt{y}}^{\sqrt{y}} e^{-(1/2)x^2}\,dx\right\} = \frac{1}{\sqrt{y}}\,e^{-(1/2)y}, \qquad 0 < y < \infty.$$

Thus

$$f_Y(y) = \begin{cases} \dfrac{1}{\sqrt{y}}\,e^{-(1/2)y}, & 0 < y < \infty, \\ 0, & -\infty < y \le 0. \end{cases}$$

In essence, it is possible to apply Theorem 1 to this problem if one considers separately each region where the transformation is one to one. We will not explain the details of the approach but refer the reader to the book by Mood and Graybill (Ref. 16). The extension of Theorem 1 to multivariate density functions is the context of the next theorem.

Theorem 2 Let $X_1, \ldots, X_n$ be continuous random variables and

$$y_i = g_i(x_1, \ldots, x_n), \qquad i = 1, 2, \ldots, n,$$

be one-to-one mappings from the n-dimensional Euclidean space into itself. Now suppose g_i and g_i^{-1} are continuous and $\partial x_i/\partial y_j$ exist and are continuous for all i and $j = 0, 1, \ldots, n$. Finally, suppose

$$J = \frac{\partial(x_1, \ldots, x_n)}{\partial(y_1, \ldots, y_n)} = \begin{vmatrix} \dfrac{\partial x_1}{\partial y_1} & \cdots & \dfrac{\partial x_1}{\partial y_n} \\ \dfrac{\partial x_2}{\partial y_1} & \cdots & \dfrac{\partial x_2}{\partial y_n} \\ \vdots & & \vdots \\ \dfrac{\partial x_n}{\partial y_1} & \cdots & \dfrac{\partial x_n}{\partial y_n} \end{vmatrix} \neq 0$$

over the range of the y_i. Then

$$f_{Y_1 \cdots Y_n}(y_1 \cdots y_n) = f_{X_1 \cdots X_n}\left(g_1^{-1}(y_1 \cdots y_n) \cdots g_n^{-1}(y_1 \cdots y_n)\right) |J|$$

The quantity J in the above theorem is called the Jacobian and is studied in most advanced calculus books.

Example 4. If X_1, X_2 is a random sample from a population with pdf $f_X(x) = e^{-x}$, $x > 0$, and zero elsewhere, find the pdf of $X_1 + X_2$.

Solution: Let

$$U = X_1 + X_2,$$

and

$$V = X_1.$$

Then the transformation $u = x_1 - x_2$ is a transformation which maps the region R_1 onto the region R_2 shown in Figure 2.1. Moreover,

$$J = \begin{vmatrix} \dfrac{\partial x_1}{\partial u} & \dfrac{\partial x_1}{\partial v} \\ \dfrac{\partial x_2}{\partial u} & \dfrac{\partial x_2}{\partial v} \end{vmatrix} = \begin{vmatrix} 0 & 1 \\ 1 & -1 \end{vmatrix} = -1.$$

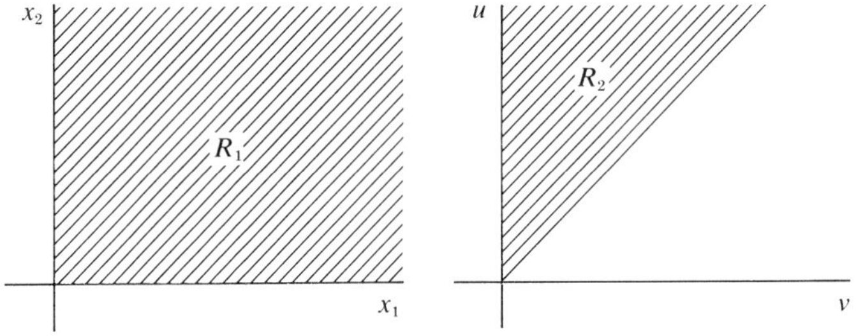

Figure 2.1 Regions of the Transformation

The transformation clearly satisfies the conditions of Theorem 2, and by noting

$$f_{X_1X_2}(x_1, x_2) = f_X(x_1)f_X(x_2) = e^{-(x_1+x_2)}, \qquad x_1, x_2 > 0,$$

we have

$$f_{VU}(v, u) = \begin{cases} e^{-u}, & \text{if } 0 < v < u, \\ 0, & \text{elsewhere.} \end{cases}$$

Thus, by Theorem 1, Section 2.7, we have

$$f_U(u) = \begin{cases} \displaystyle\int_0^u e^{-u}\,dv = ue^{-u}, & u > 0, \\ 0, & u \leq 0. \end{cases}$$

Example 5. Let X_1, X_2 be a random sample from a population with pdf $f_X(x) = 1$, $0 < x < 1$, and $f_X(x) = 0$ elsewhere. Find the pdf of Y if $Y = X_1X_2$.

Solution: Let $Y = X_1X_2$ and $V = X_2$. Then the transformation $y = x_1x_2$ and $v = x_2$ is a transformation satisfying Theorem 2. Moreover, the transformation maps R_1 onto R_2 as indicated in Figure 2.2. Thus since

$$f_{X_1X_2}(x_1, x_2) = \begin{cases} f_X(x_1)f_X(x_2) = 1, & 0 < x_1 < 1, 0 < x_2 < 1, \\ 0, & \text{elsewhere,} \end{cases}$$

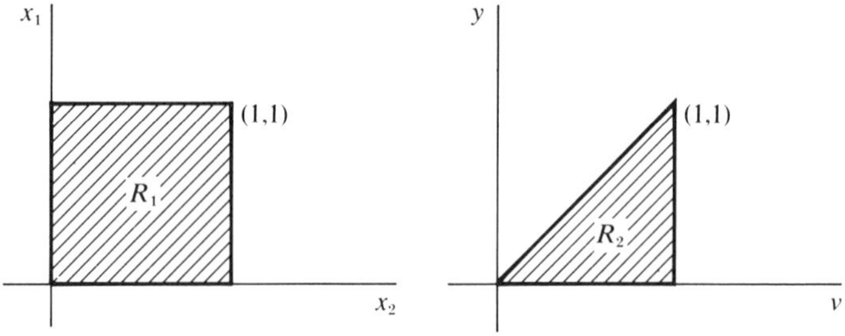

Figure 2.2 Regions of the Transformation

we have

$$f_{YV}(y, v) = 1 \circ |J| = \begin{vmatrix} \dfrac{\partial x_1}{\partial y} & \dfrac{\partial x_1}{\partial v} \\ \dfrac{\partial x_2}{\partial y} & \dfrac{\partial x_2}{\partial v} \end{vmatrix}$$

$$= \begin{vmatrix} \dfrac{1}{v} & \dfrac{-y}{v^2} \\ 0 & 1 \end{vmatrix}$$

$$= \frac{1}{v}, \qquad 0 < v < y < 1,$$

$$= 0, \qquad \text{elsewhere.}$$

Consequently,

$$f_Y(y) = \begin{cases} \displaystyle\int_y^1 \frac{dv}{v} = -\ln y = \ln y^{-1}, & 0 < y < 1, \\ 0, & \text{elsewhere.} \end{cases}$$

III
Moments and Limit Theorems

3.1 INTRODUCTION

The use of probability by the practicing engineer usually takes the form of data analysis. That is, one is normally confronted, expects to be confronted, or generates information in regard to some physical phenomenon. With this information or data it is desired to determine the probabilistic structure, or some aspect of it, of the physical phenomenon. The study of this application of the theory of probability is generally referred to as the field of statistics. Since the primary concern of this book is to aid the practicing engineer in his use of probability, our direction of study naturally turns to the statistical nature of the subject. With this in mind we will devote this chapter to the development of the tools and concepts necessary for a satisfactory probabilistic interpretation of data.

3.2 EXPECTATION

It is not unnatural for human beings, as well as engineers, to be interested in averages.[1] In fact, many problems in engineering center around such hypothetical averages (or means) as the mean breaking strength of a rod, the mean life of a fuel cell, the mean time between failures of an equipment, etc. The average most generally being considered here is the

---

[1] This statement is not meant to imply that the two sets are mutually exclusive.

arithmetic average. This average is defined as follows. Given the data (real numbers) $x_1, x_2, \ldots, x_n$ we define the arithmetic average $\bar{x}$ by

$$\bar{x} = \frac{1}{n} \sum_{i=1}^{n} x_i. \tag{1}$$

We will more commonly refer to $\bar{x}$ as a value of the sample mean,

$$\bar{X} = \frac{1}{n} \sum_{i=1}^{n} X_i.$$

That is, we think of $\bar{x}$ as an observed value of the random variable $\bar{X}$. Suppose, for example, that an automated gun fires five shots at a target and the number of hits, x, is recorded. If this experiment is repeated 10 times, and we obtain the data $x_1 = 4$, $x_2 = 3$, $x_3 = 4$, $x_4 = 3$, $x_5 = 2$, $x_6 = 4$, $x_7 = 2$, $x_8 = 3$; $x_9 = 3$, and $x_{10} = 5$, then $\bar{X}$ has the value $\bar{x}$, where

$$\bar{x} = \frac{1}{10} \sum_{i=1}^{10} x_i = 3.3, \tag{2}$$

which is the average number of hits for the observed 10 trials. Now the above example is quite trivial, but it is adequate to suggest the concept we shall shortly refer to as the expectation of a random variable. Let us consider it more closely. First we intuitively assume that a gun which is fired by an automatic device in the same manner each time does not become more accurate with practice. Thus, it would appear that the number of hits in each trial of five shots can reasonably be assumed to be random in nature. However, it would also appear that there is some fixed real number which represents the expected number or average number of hits on a trial of five shots, and that the random nature of the experiment is simply the fluctuation about this number. This number, of course, is not $\bar{x}$, since $\bar{x}$ would be different for each set of trials. However, if the experiment were repeated a large number of times, maybe several thousand times, then we might consider $\bar{x}$ as quite representative of the "true" expected number or mean number of hits in five shots. This is true, since if the experiment were repeated another several thousand trials we would not expect $\bar{x}$ to vary significantly. In this sense our notion of the "true" expected number of hits, μ, is therefore best described by

$$\mu = \lim_{n \to \infty} \frac{1}{n} \sum_{i=1}^{n} x_i. \tag{3}$$

Note that we can write (2) in the form

$$\begin{aligned}
\overline{x} &= \frac{1}{10} \sum_{i=1}^{n} x_i = \frac{1}{10}[3(4) + 4(3) + 2(2) + 1(5)] \\
&= \frac{3}{10}(4) + \frac{4}{10}(3) + \frac{2}{10}(2) + \frac{1}{10}(5) \\
&= \sum_{i=1}^{4} y_i f_n(y_i),
\end{aligned}$$

where $y_1 = 4, y_2 = 3, y_3 = 2, y_4 = 5$, and $f_n(y_i)$ is the relative frequency of occurrence of y_i in a sample of size n.

Thus it is clear that if $\overline{x}_i$ designates the *distinct* outcomes of an experiment, then (2) can be written

$$\overline{x} = \sum_{i=1}^{m} x_i f_n(x_i),$$

where m is the number of distinct values of x_i. But, from our interpretation of probability as relative frequency we see that

$$\lim_{n \to \infty} f_n(x_i) = f_X(x_i) = P[X = x_i].$$

Using this we can now rewrite (3) as

$$\mu = \sum_{i=1}^{\infty} x_i f_X(x_i),$$

where f_X is the pdf of a random variable whose values are contained in the set $\{x_1, x_2, \ldots\}$. This motivation leads us to the following definition.

Definition 1 Let X be a random variable with probability density function f_X. Then we define the expected value (or mean) of X, $E[X]$, by

$$E[X] = \begin{cases} \sum_{n=0}^{\infty} x_n f_X(x_n), & \text{if } X \text{ is discrete,} \\ \int_{-\infty}^{\infty} x f_X(x)\,dx, & \text{if } X \text{ is continuous.} \end{cases}$$

We will often use the symbol μ_X for $E[X]$. However, the notation $E[X]$ is sometimes more functional, and hence we will use the two interchangeably. Since we will be interested in the expectation of a function of X, that is, $E[g(X)]$ for various functions g, we will now extend the above definition.

Definition 2 Let X be a random variable with pdf, f_X. Let g be a function such that $g(X)$ is a discrete (continuous) random variable when X is a discrete (continuous) random variable. Then we define $E[g(X)]$ by

$$E[g(X)] = \begin{cases} \sum_{n=0}^{\infty} g(x_n) f_X(x_n), & \text{if } X \text{ is discrete,} \\ \int_{-\infty}^{\infty} g(x) f_X(x)\,dx, & \text{if } X \text{ is continuous.} \end{cases}$$

We should comment that Definition 2 is not absolutely necessary since Definition 1 gives $E[g(X)]$ in terms of the pdf, $f_{g(X)}$. This, however, is normally a difficult way to obtain $E[g(X)]$ and results in the same quantity as Definition 2.

Some important properties of the expectation operator which we will make use of are listed below:

(a) $E[aX + bY] = aE[X] + bE[Y]$, a and b are constants.

(b) $E[XY] = E[X]E[Y]$, provided X and Y are independent.

(c) $\left[E \sum_{i=1}^{n} X_i\right] = \sum_{i=1}^{n} E[X_i]$.

(d) $E[C] = C$, C a constant.

Definition 2 is extended to conditional and multivariate distributions by means of the following two definitions.

Definition 3 If $X_1, X_2, \ldots, X_n$ are random variables with joint density $f_{X_1 X_2 \cdots X_n}$, then we define $E[g(X_1 \cdots X_n)]$ by

$$E[g(X_1 \cdots X_n)] = \begin{cases} \displaystyle\sum_{k_1=0}^{\infty} \cdots \sum_{k_n=0}^{\infty} g(x_{k_1} \cdots x_{k_n}) f_{X_1 \cdots X_n}(x_{k_1} \cdots x_{k_n}), & X \text{ discrete}, \\ \displaystyle\int_{-\infty}^{\infty} \cdots \int_{-\infty}^{\infty} g(x_1 \cdots x_n) f_{X_1 \cdots X_n}(x_1 \cdots x_n)\, dx_1 \cdots dx_n, & X \text{ continuous}. \end{cases}$$

Definition 4 If X and Y are random variables with joint density, f_{XY}, we define the conditional expectation, $E[g(X) \mid y]$, by

$$E[g(X) \mid y] = \begin{cases} \displaystyle\sum_{n=0}^{\infty} g(x_n) f_{X \mid Y}(x_n \mid y) = \sum_{n=0}^{\infty} \frac{g(x_n) f_{XY}(x_n, y)}{f_Y(y)}, & X, Y \text{ discrete}, \\ \displaystyle\int_{-\infty}^{\infty} g(x) f_{X \mid Y}(x \mid y)\, dx = \int_{-\infty}^{\infty} \frac{g(x) f_{XY}(x, y)}{f_Y(y)}\, dx, & X, Y \text{ continuous}. \end{cases}$$

3.3 THE VARIANCE OF A RANDOM VARIABLE

Although the notion of average is an important one, by itself it can be very misleading. Thus, suppose five fuel cells are tested for reliability and four burn out immediately while one lasts for 1000 hours. The statement that the cells have been shown to have a 200-hour mean life (even if correct) is quite misleading. The thing that is needed is a statement in regard to how much one might expect an individual cell to vary from this mean. There are a number of ways to measure this variation about the mean. The first one that comes to mind is to simply add up the differences of the observed values and the average value and use this as a measure of variation. This, however, is of no value, since positive differences cancel negative differences and we always get zero. In the above example, for instance, if $x_1 = 1000$, $x_2 = x_3 = x_4 = x_5 = 0$, we have

$$\sum_{i=0}^{5} (x_i - 200) = 800 - 4(200) = 0.$$

An obvious way to escape this is to add up the absolute values of the differences. This, however, is undesirable because the absolute value function is difficult to work with mathematically. The difficulty can be removed and a quite reasonable, familiar measure of variation can be obtained by squaring each residual and considering the sum, that is, the quantity

$$\sum_{i=1}^{n} (x_i - \overline{x})^2.$$

This is not quite acceptable since it is not independent of the sample size. Thus large samples would lead to a large sum of squares, and hence the measure would essentially be worthless. This problem can be solved by dividing the sum by n and hence normalizing it. We are thus led to the quantity, which we designate s^2:

$$s^2 = \frac{1}{n} \sum_{i=1}^{n} (x_i - \overline{x})^2. \tag{1}$$

Since, in general, s^2 takes on different values depending on x_i, it is convenient to interpret it as a value of the random variable

$$S^2 = \frac{1}{n} \sum_{i=1}^{n} (X_i - \overline{X})^2. \tag{2}$$

The quantity S^2 in (2) is called the *sample variance.*

As we have seen s^2 is a measure of the manner in which the data varies about a value of the sample mean. By repeating the motivation used in defining $E[X]$ we would see that s^2 could be considered as an approximation to the variance about μ_X. From this point of view we would immediately be led to the following definition.

Definition 1 Let X be a random variable. Then we define the variance of X, denoted σ_X^2, by

$$\sigma_X^2 = E[(X - \mu_X)^2]. \tag{3}$$

In many instances the interest is in the positive square root of σ_X^2 and S^2. In this case we write

$$S = \sqrt{\frac{1}{n}\sum_{i=1}^{n}(X_i - \bar{X})^2} \tag{4}$$

and

$$\sigma_X = \sqrt{E[(X - \mu_X)^2]}, \tag{5}$$

and we refer to these equations as the *sample standard deviation* and *population or true standard deviation*, respectively. Alternative forms for S^2 and σ_X^2 which are easily derived and are of some importance are

$$S^2 = \frac{1}{n}\sum_{i=1}^{n} X_i^2 - \bar{X}^2 \tag{6}$$

and

$$\sigma_X^2 = E[X^2] - \{E[X]\}^2. \tag{7}$$

Example 1. Let X have the point binomial distribution defined by

$$f_X(x) = \begin{cases} p^x q^{1-x}, & x = 0, 1, \\ 0, & \text{elsewhere.} \end{cases}$$

Find μ_X and σ_X^2.

Solution:

$$\mu_X = E[X] = \sum_{x=0}^{1} x f_X(x) = f_X(1) = p,$$

$$\begin{aligned} \sigma_X^2 &= E[(X - \mu_X)^2] = E[X^2] - p^2 \\ &= \sum_{x=0}^{1} x^2 f_X(x) - p^2 \\ &= f_X(1) - p^2 \\ &= p - p^2 \\ &= p(1 - p) = pq. \end{aligned}$$

Example 2. Let

$$f_X(x) = \begin{cases} \dfrac{e^{-\lambda}\lambda^x}{x!}, & x = 0,1,2,\ldots, \\ 0, & \text{elsewhere.} \end{cases}$$

Find μ_X and σ_X^2.

Solution:

$$\begin{aligned} \mu = E[X] &= \sum_{x=0}^{\infty} \frac{xe^{-\lambda}\lambda^x}{x!} \\ &= \sum_{x=1}^{\infty} \frac{e^{-\lambda}\lambda^x}{(x-1)!} \\ &= \lambda \sum_{x=1}^{\infty} \frac{e^{-\lambda}\lambda^{x-1}}{(x-1)!} \\ &= \lambda \sum_{x=0}^{\infty} \frac{e^{-\lambda}\lambda^x}{x!} \\ &= \lambda \sum_{x=0}^{\infty} f_X(x) \\ &= \lambda(1) = \lambda. \end{aligned}$$

By using the trick $E[X^2] = E[X(X-1)] + E[X]$, one can show in the same fashion that $\sigma_X^2 = \lambda$. When X has the pdf of this example, it is said to have Poisson distribution. The density was first defined by Poisson in connection with the problem of determining the pdf for death from the kick of a horse.

Example 3. Let $X_1, \ldots, X_n$ be a random sample of size n with pdf f_X. Show that

(a) $E[\overline{X}] = E[X] = \mu_X$, i.e., $\mu_X = \mu_{\overline{X}}$;

(b) $E[(\overline{X} - \mu)^2] = \dfrac{1}{n} E[(X - \mu)^2] = \dfrac{\sigma_X^2}{n}$, i.e., $\sigma_{\overline{X}}^2 = \dfrac{\sigma_X^2}{n}$.

Solution:

$$\text{(a) } E[\overline{X}] = E\left[\frac{1}{n}\sum_{i=1}^{n} x_i\right]$$

$$= \frac{1}{n}\sum_{i=1}^{n} E[X_i]$$

$$= \frac{1}{n}\{E[X_1] + \cdots + E[X_n]\}$$

$$= \frac{1}{n}\{\mu_X + \cdots + \mu_X\}$$

$$= \frac{n\mu_X}{n} = \mu_X.$$

$$\text{(b) } E[(\overline{X} - \mu_X)^2] = E\left[\left(\frac{1}{n}\sum_{i=1}^{n} X_i - \mu_X\right)^2\right]$$

$$= E\left[\left\{\frac{1}{n}\sum_{i=1}^{n}(X_i - \mu_X)\right\}^2\right]$$

$$= \frac{1}{n^2} E\left[\left\{\sum_{i=1}^{n}(X_i - \mu_X)\right\}^2\right].$$

Since

$$E[(X_i - \mu_X)(X_j - \mu_X)] = E[(X_i - \mu_X)]E[(X_j - \mu_X)] = 0,$$

when $i \neq j$ by the independence of X_i and X_j, then

$$E[(\overline{X} - \mu_X)^2] = \frac{1}{n^2}\sum_{i=1}^{n} E(X_i - \mu_X)^2$$

$$= \frac{n\sigma_X^2}{n^2} = \frac{\sigma_X^2}{n}.$$

The results of this example are quite fundamental and we shall use them frequently.

Example 4. Let X be distributed normal with parameters a and b, that is,

$$f_X(x) = \frac{1}{b\sqrt{2\pi}} \exp\left[-\frac{1}{2}\left(\frac{x-a}{b}\right)^2\right] \quad -\infty < x < \infty.$$

Find μ_X and σ_X^2.

Solution: It is not difficult to determine μ_X and σ_X^2 from their definitions. However, in order to point out a useful result, we will take an easier path. It can be shown that if f_X is a pdf with a finite mean μ_X and if $f_X(x)$ is symmetric about some point λ, that is, $f_X(\lambda - x) = f_X(\lambda + x)$ for every x, then $\lambda = \mu_X$. Now

$$E[X] = \frac{1}{b\sqrt{2\pi}} \int_{-\infty}^{\infty} x \exp\left[-\frac{1}{2}\left(\frac{x-a}{b}\right)^2\right] dx$$

exists, and clearly $f_X(a - x) = f_X(a + x)$ for every x. Hence

$$\mu_X = E[X] = a.$$

In order to determine σ_X^2 one needs to make use of the well-known gamma function, but we will not do it here. The result, however, is $\sigma_X^2 = b^2$. For this reason the pdf of this example is most often written

$$f_X(x) = \frac{1}{\sigma_X\sqrt{2\pi}} \exp\left[-\frac{1}{2}\left(\frac{x-\mu_X}{\sigma_X}\right)^2\right], \qquad -\infty < x < \infty,$$

and X is said to be distributed normal with mean μ_X and variance σ_X^2 or $X \sim N(\mu_X, \sigma_X^2)$.

3.4 HIGHER-ORDER MOMENTS

In the above we have discussed two parameters, μ_X and σ_X^2, which to some extent describe a probability density function. The parameter μ_X gives us a number which represents the average value of the random variable X while the number σ_X^2 is a measure of how the values of X cluster about μ_X. An equally important property of μ_X and σ_X^2 is that in many cases a probability model which is completely characterized by one or both of these quantities can be reasonably assumed for a physical problem. For example, the probability density functions of Examples 1, 2, and 4 are all very important as realistic probability models. The first two are completely determined if μ_X is known, while the pdf of Example 4 is completely specified by μ_X and σ_X^2.

The quantities μ_X and σ_X^2 are commonly also referred to as the first and second moments. It is convenient to extend this notion in the following way.

Definition 1 If X is a random variable and b is a real number, then we define the rth moment about b, $\mu_r(b)$, by

$$\mu_r(b) = E[(X - b)^r]. \tag{1}$$

In particular,

$$\mu_X = \mu_1(0) \qquad \text{and} \qquad \sigma_X^2 = \mu_2(\mu_X).$$

The quantities defined by (1) are of some practical importance since they essentially determine the probability density function and, as we shall see, can be measured to some degree from physical data. The most important moments are, of course, the first two. However, some work of significance has been done with higher-order moments. Pearson, for example, in some classical work has used the first four moments for estimating probability density functions from random data quite successfully. See Kendall and Stuart (Ref. 12).

In Definition 1 we have defined what might be considered the rth "theoretical" moment about b. Previously, we have discussed some sample moments, $\overline{X}$ and S^2. The notion of sample moments is extended, for $b = 0$, to higher-order sample moments as follows.

Definition 2 Let $X_1, \ldots, X_n$ be a random sample of size n. Then we define the rth sample moment about zero, M_r, by

$$M_r = \frac{1}{n} \sum_{k=1}^{n} X_k^r. \tag{2}$$

Note that

$$\overline{X} = M_1$$

and

$$S^2 = M^2 - M_1^2.$$

It is also worth noting that

$$
\begin{aligned}
E[M_r] &= \frac{1}{n}\sum_{k=1}^{n} E[X_k^r] \\
&= \frac{1}{n}\sum_{k=1}^{n} \mu_r(0) = \mu_r(0).
\end{aligned} \tag{3}
$$

Clearly, a special case of (3) is $E[\overline{X}] = \mu_X$.

As we will see in later chapters, the primary value of sample moments will be the information they convey in regard to the theoretical moments, and hence the probability density function.

The notion of moments extends to multivariate distributions. However, the only ones we will have need for in the multivariate case are $E[(X - \mu_X)(Y - \mu_Y)]$ and $E[XY]$, which are called the covariance and cross-correlation, respectively. The corresponding sample moments are given by

$$
S_{XY} = \frac{1}{n}\sum_{i=1}^{n} (X_i - \overline{X})(Y_i - \overline{Y})
$$

and

$$
r_{XY} = \frac{1}{n}\sum_{i=1}^{n} X_i Y_i
$$

for random samples $(X_1, \ldots, X_n)$ and $(Y_1, \ldots, Y_n)$. Note that if X and Y are independent, the covariance of X and Y is zero; that is,

$$
\begin{aligned}
E[(X - \mu_X)(Y - \mu_Y)] &= E[X - \mu_X]E[Y - \mu_Y] \\
&= 0.
\end{aligned}
$$

3.5 THE MOMENT-GENERATING FUNCTION

In Section 3.4 we pointed out that a probability density function can for practical purposes be characterized by its moments. In fact, as also indicated above, in some instances the first two moments may be of more interest than the pdf.

In this section we introduce a function which is quite useful in determining the moments of a random variable X and the manner in which they characterize the pdf.

Definition 1 Let X be a random variable and t a real number. Then if $E[e^{Xt}]$ exists for all t in some interval about zero, we define the moment-generating function $M_X(t)$ by

$$M_X(t) = E[e^{Xt}]. \tag{1}$$

Example 1. Find $M_X(t)$ if

$$f_X(x) = \begin{cases} \binom{n}{x} p^x q^{n-x}, & x = 0,1,2,\ldots,n, \\ 0, & \text{elsewhere.} \end{cases}$$

Solution:

$$\begin{aligned} M_X(t) = E[e^{Xt}] &= \sum_{x=0}^{n} e^{tx}\binom{n}{x} p^x q^{n-x} \\ &= \sum_{x=0}^{n} \binom{n}{x} (e^t p)^x q^{n-x} \\ &= (e^t p + q)^n. \end{aligned}$$

Example 2. Find $M_X(t)$ if $X \sim N(\mu, \sigma^2)$, that is, if

$$f_X(x) = \frac{1}{\sigma\sqrt{2\pi}} \exp\left[-\frac{1}{2}\left(\frac{x-\mu}{\sigma}\right)^2\right], \qquad -\infty < x < \infty.$$

Solution:

$$\begin{aligned} E[e^{Xt}] &= \frac{1}{\sigma\sqrt{2\pi}} \int_{-\infty}^{\infty} e^{xt} \exp\left[-\frac{1}{2}\left(\frac{x-\mu}{\sigma}\right)^2\right] dx \\ &= \frac{1}{\sigma\sqrt{2\pi}} \int_{-\infty}^{\infty} \exp\left[xt - \frac{1}{2}\left(\frac{x-\mu}{\sigma}\right)^2\right] dx \\ &= e^{\mu t + (\sigma^2 t^2)/2} \frac{1}{\sigma\sqrt{2\pi}} \int_{-\infty}^{\infty} \exp\left[-\frac{1}{2}\left(\frac{x-\mu-t\sigma^2}{\sigma}\right)^2\right] dx \\ &= e^{\mu t + (\sigma^2 t^2)/2}. \end{aligned}$$

The function $M_X(t)$ can be used to determine the moments about zero of the random variable X by the following considerations. When $M_X(t)$ exists we have

$$\frac{d^k}{dt^k} E[e^{Xt}] = \begin{cases} \displaystyle\int_{-\infty}^{\infty} \frac{d^k}{dt^k} e^{xt} f_X(x)\,dx = \int_{-\infty}^{\infty} x^k e^{xt} f_X(x), & X \text{ continuous}, \\ \displaystyle\sum_{n=0}^{\infty} \frac{d^k}{dt^k} e^{x_n t} f_X(x_n) = \sum_{n=0}^{\infty} x_n^k e^{x_n t} f_X(x), & X \text{ discrete}. \end{cases}$$

Thus

$$\left. \frac{d^k}{dt^k} M_X(t) \right|_{t=0} = E[X^k]. \tag{2}$$

The result (2) can also be seen by noting that

$$e^{xt} = \sum_{n=0}^{\infty} x^k \frac{t^k}{k!},$$

and hence

$$E[e^{Xt}] = \sum_{n=0}^{\infty} E[X^k] \frac{t^k}{k!} = 1 + \mu_1(0)t + \mu_2(0) \frac{t^2}{2!} + \cdots,$$

so that (2) again follows.

Equation (2) represents a very important property of the moment-generating function. However, the following two theorems are probably the most important properties of this function.

Theorem 1 Let X and Y be two random variables with pdf's f_X and f_Y. Now suppose $M_X(t)$ and $M_Y(t)$ exist and

$$M_X(t) = M_Y(t)$$

for all t in some interval about the origin. Then $f_X = f_Y$.[1]

---

[1] If X and Y are continuous, then it can happen that $f_X(x) \neq f_Y(x)$ for certain x's. However, it can be shown that these values occur with a probability of zero.

Theorem 2 Let $X_1, \ldots X_n$ be independent random variables with moment-generating functions $M_{X_1}(t)$, $M_{X_2}(t)$, $\ldots, M_{X_n}(t)$. Then if

$$Y = \sum_{i=1}^{n} X_i,$$

we have

$$M_Y(t) = M_{X_1}(t) M_{X_2}(t) \cdots M_{X_n}(t). \tag{3}$$

Note that if $X_1, \ldots, X_n$ is a random sample, then

$$M_{X_1}(t) = M_{X_2}(t) = \cdots = M_{X_n}(t)$$

and

$$M_Y(t) = [M_X(t)]^n. \tag{4}$$

The value of Theorems 1 and 2 will be demonstrated in the following example.

Example 3. Let $X_1, \ldots, X_n$ be a random sample from a $N(\mu, \sigma^2)$ population. Find the pdf of $\overline{X}$.

Solution: First, note that for any constants a and b,

$$\begin{aligned} M_{aX+b}(t) &= E[e^{(aX+b)t}] \\ &= E[e^{atX} e^{bt}] \\ &= e^{bt} E[e^{atX}] \\ &= e^{bt} M_X(at). \end{aligned} \tag{5}$$

Thus

$$M_{\overline{X}}(t) = M_Y\left(\frac{t}{n}\right),$$

where

$$Y = \sum_{i=1}^{n} X_i.$$

But

$$M_Y\left(\frac{t}{n}\right) = \left[M_X\left(\frac{t}{n}\right)\right]^n$$

by (5), and

$$\begin{aligned} \left[M_X\left(\frac{t}{n}\right)\right]^n &= \left\{\exp\left[\mu\left(\frac{t}{n}\right) + \frac{1}{2}\sigma^2\left(\frac{t}{n}\right)^2\right]\right\}^n \\ &= \exp\left(\mu t + \frac{1}{2}\frac{\sigma^2}{n}t^2\right). \end{aligned} \tag{6}$$

Now, by Example 2, (6) is the moment-generating function of a normal distribution with mean μ and variance σ^2/n. Hence, from Theorem 1, $\overline{X}$ has a normal distribution with mean μ and variance σ^2/n, that is, $\overline{X} \sim N(\mu, \sigma^2/n)$.

Example 3 illustrates a general method for finding the pdf of a sum of random variables. In Example 3, however, we have depended on our ability to recognize the moment-generating function of $M_{\overline{X}}(t)$ as one associated with a familiar pdf. This is in general not necessary if one makes use of the Laplace transforms. For example, the one-sided Laplace transform L of a function f is defined by

$$L[f(x)] = \int_0^\infty e^{-sx} f(x)\, dx = F(s)$$

and hence, if $f_X(t) = 0$ when $t < 0$, $L[f_X(x)] = M_X(t)$, where $t = -s$. Thus

$$f_X(x) = L^{-1} M_X(-s). \tag{7}$$

Equation (7) simply says that if we know the moment-generating function $M_X(t)$, then, by letting $t = -s$, M_X is a function of s and $f_X(x)$ is simply the inverse Laplace transform of this function. The same result applies to functions which do not satisfy the condition $f_X(x) = 0$ for $t < 0$, but in this case the two-sided Laplace transform must be used.

Example 4. Let X_1, X_2, X_3 be a random sample with pdf

$$f_X(x) = \begin{cases} 1, & 0 < x < 1, \\ 0, & \text{elsewhere.} \end{cases}$$

Find the pdf of $Y = X_1 + X_2 + X_3$.

Solution:

$$M_X(t) = \int_0^1 e^{tx}\, dx = \frac{1}{t}(e^t - 1).$$

Thus

$$M_Y(t) = \frac{1}{t^3}(e^t - 1)^3$$

or

$$M_Y(t) = \frac{1}{t^3}\,[e^{3t} - 3e^{2t} + 3e^t - 1]$$

and

$$M_Y(-s) = \frac{-1}{s^3}\,[e^{-3s} - 3e^{-2s} + 3e^{-s} - 1],$$

so that

$$\begin{aligned}
f_Y(t) &= L^{-1}\left\{\frac{1}{s^3}\,(-e^{-3s} + 3e^{-2s} - 3e^{-s} + 1)\right\} \\
&= -L^{-1}\left\{\frac{e^{-3s}}{s^3}\right\} + 3L^{-1}\left\{\frac{e^{-2s}}{s^3}\right\} - 3L^{-1}\left\{\frac{e^{-s}}{s^3}\right\} + L^{-1}\left\{\frac{1}{s^3}\right\} \\
&= \tfrac{1}{2}t^2, && 0 < t < 1, \\
&= \tfrac{1}{2}t^2 - \tfrac{3}{2}(t-1)^2, && 1 \le t < 2, \\
&= \tfrac{1}{2}t^2 - \tfrac{3}{2}(t-1)^2 + \tfrac{3}{2}(t-2)^2, && 2 \le t < 3, \\
&= 0, && \text{elsewhere.}
\end{aligned}$$

For the reader who is familiar with the Laplace transforms the above inverse transforms follow by inspection. For others, the results are easily obtained from a table of Laplace transforms and their inverses. See Erdélyi (Ref. 8).

3.6 LAWS OF LARGE NUMBERS

In the previous sections of this chapter we have studied some aspects of the relationship between the pdf and the moments of a random variable X. In this section we will explore this relationship further by considering the behavior of the sample mean as the sample size increases. Intuitively, we expect $\overline{x}$ to be quite close to μ_X if n is large. We will now see how it is possible to put this concept in a precise form and hence determine the exact meaning of such nebulous words as "close" and "large."

Theorem 1 Let X be a random variable with pdf f_X such that $E[X]$ is finite and

$$f_X(x) = 0 \qquad \text{for } x < 0.$$

Then

$$P[X \geq a] \leq \frac{E[X]}{a} \qquad \text{for any } a > 0. \tag{1}$$

Proof: We include a proof of the above since it is quite simple. It is as follows.

$$E[X] = \int_0^\infty x f_X(x)\,dx \geq \int_a^\infty x f_X(x)\,dx$$

$$\geq a \int_a^\infty f_X(x)\,dx = aP[X \geq a].$$

The proof is entirely similar if X is a discrete random variable.

Corollary 1 Let X be a random variable with finite variance σ_X^2 and mean μ_X. Then, for $k > 0$,

$$P[\,|X - \mu_X| > k\sigma_X] \leq \frac{1}{k^2} \tag{2}$$

Proof: By Theorem 1

$$P[\,|X - \mu_X| > k\sigma_X]$$

$$= P[\,|X - \mu_X|^2 > (k\sigma_X)^2] \leq \frac{E[\,|X - \mu_X|^2]}{(k\sigma_X)^2}.$$

Thus

$$P[\,|X - \mu_X| > k\sigma_X] \leq \frac{E[(X - \mu_X)^2]}{(k\sigma_X)^2}$$

$$= \frac{\sigma_X^2}{k^2\sigma_X^2} = \frac{1}{k^2}.$$

Corollary 2 Let $X_1, \ldots, X_n$ be a random sample from a population with mean μ_X and finite variance σ_X^2. Then

$$P\left[\,|\overline{X}_n - \mu| > k\frac{\sigma}{\sqrt{n}}\right] \leq \frac{1}{k^2}. \tag{3}$$

Proof: Substitute $\sigma_{\overline{X}}$ for σ_X in (2) and note that

$$\sigma_{\overline{X}_n}^2 = \frac{\sigma_X^2}{n}.$$

The result follows.

The inequalities in (2) and (3) are called Tchebycheff inequalities and we will now demonstrate some practical importance of them. It should be noticed that a specific density function for X is not assumed in these results.

Example 1. Determine n sufficiently large that "$|\overline{X} - \mu_X| < .1$ at least 99 percent of the time."

Assume $\sigma_X^2 = 4$.

Solution: By "$|\overline{X} - \mu_X| < .1$ at least 99 percent of the time" we mean determine n so that

$$P[\,|\overline{X}_n - \mu| > .1] \leq .01.$$

Thus in (3) we have

$$\frac{1}{k^2} = .01$$

and

$$k\,\frac{\sigma_X}{\sqrt{n}} = \frac{2k}{\sqrt{n}} = .1.$$

Solving for k and n we have $k = 10$ and $n = 40{,}000$.

The following theorem, which is referred to as the weak law of large numbers, guarantees us that such n as determined in Example 1 always exists.

Theorem 2 (Weak Law of Large Numbers). Let $X_1, X_2, \ldots, X_n$ be a random sample with mean μ_X. Then $\overline{X}_n$ converges in probability to μ_X; that is, for any $\epsilon > 0$,

$$\lim_{n \to \infty} P[\,|\overline{X}_n - \mu_X| > \epsilon] = 0. \tag{4}$$

Note that if σ_X^2 exists, then (4) is an immediate consequence of (3). This is seen by restating (4) in terms of the definition of a limit. That is, the statement (4) says that given any $\epsilon > 0$ and

$\delta > 0$, an N can be found such that

$$P[\,|\overline{X}_n - \mu_X| > \epsilon] \leq \delta, \qquad \text{when } n > N.$$

The theorem then follows simply by determining N from (3). Letting $k^2 = 1/\delta$ we see that $N = \sigma_X^2/\delta\epsilon^2$. We should emphasize, however, that even when σ_X^2 is not finite, (4) holds, and hence for sufficiently large n we expect $\overline{x}$ to be close to μ_X.

Statements such as these do not guarantee that a particular $\overline{x}$ will be within the desired proximity of μ_X no matter how large n may be. They simply guarantee that in "most" instances this will be the case, where "most" is formally defined by equations such as (3). Thus, in Example 1 "most" means at least 99 percent of the time or in at least 99 percent of such samples.

The above theorems show that $\overline{x}_n$ can be expected to approximate μ_X with any desired degree of accuracy provided n is sufficiently large. A similar result can be obtained for the sample variance. This is the context of the next theorem.

Theorem 3 Let $X_1, \ldots, X_n$ be a random sample from a population with finite fourth moment, $E[X^4]$. Then for any $\epsilon > 0$,

$$\lim_{n \to \infty} P[\,|S_n^2 - \sigma_X^2| > \epsilon] = 0, \tag{5}$$

where

$$S_n^2 = \frac{1}{n} \sum_{i=1}^{n} (X_i - \overline{X})^2.$$

IV
Important Probability Density Functions

4.1 INTRODUCTION

We have discussed probability density functions and their significance. In this chapter we will discuss a number of the more important ones and develop some of their more pertinent properties. The notation $X \sim f_X$, which is read "X is distributed f_X" or "X has the probability density function f_X," will be used on occasions.

4.2 THE POINT BINOMIAL DISTRIBUTION

The point binomial distribution has been motivated and discussed in previous chapters, but we will list it again here for completeness.

Let

$$f_X(x) = \begin{cases} p^x q^{1-x}, & x = 0,1, \\ 0, & \text{elsewhere,} \end{cases} \tag{1}$$

where $q = 1 - p$.

Then we say X has a point binomial distribution. The following properties are easily shown and in fact were established in Chapter II.

$$\mu_X = E[X] = p, \tag{2}$$

$$\sigma_X^2 = E[(X - \mu_X)^2] = pq, \tag{3}$$

$$M_X(t) = E[e^{Xt}] = (pe^t + q). \tag{4}$$

Also, by the weak law of large numbers for a random sample of size n, we have

$$\bar{X}_n = \frac{1}{n}\sum_{i=1}^{n} X_i \to p \tag{5}$$

in probability.

4.3 THE BINOMIAL DISTRIBUTION

Let

$$f_X(x) = \begin{cases} \binom{n}{x} p^x q^{n-x}, & x = 0,1,2,\ldots,n, \\ 0, & \text{elsewhere,} \end{cases} \tag{1}$$

where $q = 1 - p$.

Then X is said to have a binomial distribution. Note that the binomial distribution is a generalization of the point binomial distribution and that it possesses two parameters, p and n. In Example 6, Section 2.4, we saw that $f_X(x)$ is the probability of exactly x successes in n trials if

(a) one of two mutually exclusive outcomes occurs at each trial,
(b) the outcomes of each trial are independent, and
(c) the probability of a success on a single trial, p, remains fixed from trial to trial.

Example 1. Suppose the probability that a retro rocket will be defective after launching and two weeks in space is .05. How many such rockets should a spaceship have if it is desired for the probability that at least one will fire after such a trip be at least .999?

Solution: Assume that the rockets fire independently. Using the fact that p is fixed and is the same for each rocket, we have

$$P[X = x] = \binom{n}{x} p^x q^{n-x},$$

where x is the number of rockets out of n rockets which are not defective. Thus X has a binomial distribution and

$$P[X \geq 1] = 1 - F_X(0) = 1 - q^n.$$

We desire that n be chosen such that

$$1 - q^n \geq .999$$

or

$$(.05)^n \leq .001$$

or

$$n \geq \frac{\log(.001)}{\log(.05)} \cong 2.3.$$

Thus three rockets will suffice.

To find μ_X and σ_X^2 let us first determine the moment-generating function, M_X:

$$\begin{aligned} M_X(t) &= \sum_{x=0}^{n} \binom{n}{x} e^{tx} p^x q^{n-x} \\ &= \sum_{x=0}^{n} \binom{n}{x} (e^t p)^x q^{n-x} \\ &= (e^t p + q)^n. \qquad (2) \end{aligned}$$

Thus

$$\begin{aligned} \mu_X = \frac{d}{dt}[(e^t p + q)^n]\big|_{t=0} &= e^t np(e^t p + q)^{n-1}\big|_{t=0} \\ &= np \qquad (3) \end{aligned}$$

and

$$\begin{aligned} E[X^2] &= \frac{d^2}{dt^2}[(e^t p + q)^n]\big|_{t=0} \\ &= e^{2t} n(n-1)p^2(e^t p + q)^{n-2}\big|_{t=0} + e^t np(e^t p + q)^{n-1}\big|_{t=0} \\ &= n(n-1)p^2 + np, \end{aligned}$$

so that

$$\begin{aligned}\sigma_X^2 &= E[X^2] - \{E[X]\}^2 \\ &= npq. \end{aligned} \tag{4}$$

We conclude this section by showing that if $X_1, X_2, \ldots, X_n$ is a random sample of size n from a point binomial distribution, then $Y = n\overline{X}$ has a binominal distribution. In the process we will establish a general method for deriving densities of functions of discrete random variables.

Now

$$Y = n\overline{X} = \sum_{i=1}^{n} X_i,$$

where the possible values of X_i are 0 and 1. Thus the range of $n\overline{X}$ (that is, set of possible values of $n\overline{X}$) is $\{0,1,\ldots,n\}$. The random variable Y will have the value y if and only if X_i has the value 1, exactly y times. But the probability of y 1's and $n - y$ 0's is $p^y q^{n-y}$ for a particular order. However, all orders are equally likely, and we have previously shown there are $\binom{n}{y}$ distinct ways of arranging n 1's and $n - y$ 0's. Thus

$$\begin{aligned} f_Y(y) = P[Y = y] &= \overbrace{p^y q^{n-y} + \cdots + p^y q^{n-y}}^{\binom{n}{y} \text{ terms}} \\ &= \binom{n}{y} p^y q^{n-y}, \qquad y = 0,1,\ldots,n, \end{aligned}$$

as claimed.

4.4 THE NEGATIVE BINOMIAL DISTRIBUTION

Consider now the following problem. Suppose a certain mathematician has a memoryless random "epsilon giver" with an infinite supply of epsilons. Suppose further that the probability of obtaining a "small" epsilon from the "epsilon giver" is 0.2. What is the probability that one will have to be given at least five epsilons in order to get three "small" ones?

Solution: Let x = number of epsilons given, to obtain three "small" ones. Then

$$P[X \geq 5] = \sum_{k=5}^{\infty} P[X = k].$$

Thus we need to determine the probability that exactly x epsilons will have to be given in order to obtain three "small" ones.

Now, if $x < 3$, it is clear that $P[X = x] = 0$. However, if $x \geq 3$, we note that the probability the last epsilon is "small" is .2, and the probability that exactly 2 of the $x - 1$ other epsilons are small is given by the binomial density function and is $\binom{x-1}{2}(.2)^2(.8)^{x-3}$. But, owing to the lack of memory of the machine, epsilon giving is an independent phenomenon. Thus

$$P[X = x] = .2\left[\binom{x-1}{2}(.2)^2(.8)^{x-3}\right] = \binom{x-1}{2}(.2)^3(.8)^{x-3}.$$

Therefore,

$$\begin{aligned} P[X \geq 5] &= \sum_{k=5}^{\infty}\binom{k-1}{2}(.2)^3(.8)^{k-3} \\ &= 1 - \sum_{k=3}^{4}\binom{k-1}{2}(.2)^3(.8)^{k-3} \\ &= 1 - (.008 + .192) = 0.8. \end{aligned}$$

The above is an example of the following general type of probability problem. Suppose the assumptions necessary for a binomial distribution are satisfied; that is, each trial results in one of two mutually exclusive outcomes S and F, the outcome of each trial is independent of other trials, and the probability of S and F remains fixed at p and q, respectively. What is the probability that x trials will be required to produce r of the outcomes S (or r successes)? If x is the number of required trials we have

$$P[X = x] = f_X(x) = \begin{cases} \binom{x-1}{r-1} p^r q^{x-r}, & x = r, r+1, \ldots, \\ 0, & \text{elsewhere.} \end{cases} \tag{1}$$

The pdf $f_X(x)$ in (1) is called the negative binomial distribution and can be written in the alternative form

$$f_X(x) = \begin{cases} \binom{x+r-1}{r-1} p^r q^x, & x = 0,1,2, \ldots, \\ 0, & \text{elsewhere,} \end{cases} \tag{2}$$

where in this case $x + r$, rather than x, is the number of trials needed to produce r successes.

Using (1) we have

$$\begin{aligned} E[e^{tx}] &= \sum_{x=r}^{\infty} \binom{x-1}{r-1} p^r q^{x-r} e^{tx} \\ &= p^r \sum_{x=0}^{\infty} \binom{x+r-1}{r-1} q^x e^{tx+rt} \\ &= p^r e^{rt} \sum_{x=0}^{\infty} \binom{x+r-1}{r-1} (e^t q)^x \\ &= e^{rt} \left(\frac{p}{1 - e^t q} \right)^r, \end{aligned} \tag{3}$$

since, in general, it is well known that

$$\sum_{x=0}^{\infty} \binom{x+r-1}{r-1} \alpha^x = \left(\frac{1}{1-\alpha} \right)^r$$

if $|\alpha| < 1$. Thus

$$\begin{aligned} \mu_X = \frac{d}{dt} M_X(t) \bigg|_{t=0} &= \frac{rq}{p} + r \\ &= r\left(\frac{q+p}{p}\right) = \frac{r}{p} \end{aligned} \tag{4}$$

and

$$\sigma_X^2 = \frac{d^2}{dt^2} M_X(t) \mid_{t=0} - \left(\frac{r}{p}\right)^2 = \frac{rq}{p^2}. \tag{5}$$

Note that if $x + r$ is the number of trials as in (2) that

$$M_X(t) = \sum_{x=0}^{\infty} \binom{x + r - 1}{r - 1} p^r q^{x-r} e^{tx}$$

$$= \left(\frac{p}{1 - e^t q}\right)^r \tag{6}$$

and hence

$$\mu_X = \frac{rq}{p} \tag{7}$$

and

$$\sigma_X^2 = \frac{rq}{p^2}. \tag{8}$$

This could, of course, be obtained from (3), (4), and (5) as well. For example, since by (4)

$$E[X + r] = \frac{rq}{p} + r$$

and since

$$E[X + r] = E[X] + E[r] = E[X] + r,$$

we see that

$$E[X] = \frac{rq}{p}.$$

4.5 THE HYPERGEOMETRIC DISTRIBUTION

Suppose 10 television sets are selected at random from each lot of 100 television sets and inspected thoroughly, as part of a manufacturer's quality-control program. Suppose, further, that if more than 2 sets in the 10 are classified as defective, then the price of the lot is reduced and it is sold as an "off-brand" model to a customer who prefers to handle so-called offbrand models. If in a certain lot of 100, 10 percent of the sets are defective, what is the probability they will be sold as "offbrands"?

Solution: Let x = the number of defectives found in a sample of 10.

Consider the process of selecting and inspecting each of the 10 televisions as the outcome of an experiment and let a nondefective set be termed a success and a defective set termed a failure. Then we are asking the question of three or more failures in 10 trials, which is very much like the general problem solved by the binomial distribution. However, there is a basic difference: that the probabilities p and q are not constant from trial to trial. Thus, for example, if the first set inspected is found to be defective, then the probability the second will be defective is $\frac{9}{99}$, while if the second one is not defective, the probability that the third is defective is $\frac{9}{98}$, etc. To solve the problem we note that x defectives can be drawn from 10 defectives in $\binom{10}{x}$ ways, while $10 - x$ nondefectives can be selected from 90 nondefectives in $\binom{90}{10-x}$. Thus, the number of ways one can select x defectives and $10 - x$ nondefectives is $\binom{10}{x}\binom{90}{10-x}$. It then follows that

$$P[X = x] = \frac{\binom{10}{x}\binom{90}{10-x}}{\binom{100}{10}}, \qquad x = 0,1,\ldots,10,$$

$$= 0, \qquad \text{elsewhere.} \qquad (1)$$

Thus

$$P[X \geq 3] = 1 - F_X(2) = 1 - \sum_{x=0}^{1} \frac{\binom{10}{x}\binom{90}{10-x}}{\binom{100}{10}}$$

$$= 1 - \frac{\binom{90}{10}}{\binom{100}{10}} - \frac{\binom{10}{1}\binom{90}{9}}{\binom{100}{10}} - \frac{\binom{10}{2}\binom{90}{8}}{\binom{100}{10}}$$

$$= .06.$$

The distribution defined by (1) is a special case of the hypergeometric distribution and is defined, in general, by the density function

$$f_X(x) = \begin{cases} \dfrac{\binom{m}{x}\binom{k}{n-x}}{\binom{m+k}{n}}, & x = 0,1,\ldots,n, \\ 0, & \text{elsewhere.} \end{cases} \tag{2}$$

It should be noted that if n is small compared to $m + k$, then p and q from trial to trial do not change much, and hence the binomial distribution would be expected to approximate the hypergeometric distribution in that case.

If we use that approximation in this problem, we have

$$P[X \geq 3] \cong 1 - \sum_{x=0}^{2} \binom{10}{x} (.1)^x (.9)^{10-x}$$

$$= .07.$$

4.6 THE POISSON DISTRIBUTION

Let

$$f_X(x) = \begin{cases} \dfrac{e^{-m} m^x}{x!}, & x = 0,1,2,\ldots, m > 0, \\ 0, & \text{elsewhere.} \end{cases} \tag{1}$$

Then we say X has a Poisson distribution.

The Poisson distribution is of some importance in many physical problems. It has been found to be a satisfactory probability model in such fields as reliability and communication.

From (1) we have

$$
\begin{aligned}
E[e^{Xt}] &= \sum_{x=0}^{\infty} \frac{e^{-m}m^x}{x!} e^{xt} \\
&= \sum_{x=0}^{\infty} \frac{e^{-m}(me^t)^x}{x!} \\
&= e^{-m}e^{me^t} \\
&= e^{m(e^t-1)}. \qquad (2)
\end{aligned}
$$

From (2) one can obtain μ and σ^2 quite easily. However, recall that in Example 2, Section 3.3, it was shown that $\mu = \sigma^2 = m$. Since $\mu = m$, (1) is often written in the form

$$
f_X(x) = \begin{cases} \dfrac{e^{-\mu}\mu^x}{x!}, & x = 0,1,2,\ldots, \\ 0, & \text{elsewhere.} \end{cases} \qquad (3)
$$

There are a number of motivations for the probability model (1). We will now give two. The first will be a physical motivation and the second a mathematical motivation.

First Motivation Let T be a fixed time period and assume that the occurrence of a certain event, E, in any time interval $S_1 < T$ is independent of its occurrence in any other nonoverlapping time interval $S_2 < T$. Further, suppose the probability of this event occurring in a time interval of length Δt is proportional to Δt and that for Δt sufficiently small the probability that E occurs more than once in the interval, which we will now call for convenience Δt, is negligible. Then

$$
P[E \text{ occurs within } \Delta t] = k\,\Delta t, \qquad \text{for some } k.
$$

Now consider the problem of determining the probability, $P_x(t)$ that E will occur exactly x times within a time t. First we note that

$$
\begin{aligned}
P_0(\Delta t) &= 1 - P_1(\Delta t) \\
&= 1 - k\,\Delta t, \qquad \text{for } \Delta t \text{ sufficiently small.}
\end{aligned}
$$

But for small Δt and $t + \Delta t < T$ we have

$$\begin{aligned} P_x(t + \Delta t) &= P[E \text{ occurs exactly } x \text{ times within time } t \text{ and } 0 \text{ times in } \Delta t] + P[E \text{ occurs } x - 1 \text{ times within time } t \text{ and } 1 \text{ time in time } \Delta t] \\ &= P[E \text{ occurs } x \text{ times within time } t]\, P[E \text{ occurs } 0 \text{ times in } \Delta t] + P[E \text{ occurs } x - 1 \text{ times within time } t]\, P[E \text{ occurs } 1 \text{ time in } \Delta t] \\ &= P_x(t)[1 - k\,\Delta t] + P_{x-1}(t)\,k\,\Delta t \end{aligned}$$

or

$$\frac{P_x(t + \Delta t) - P_x(t)}{\Delta t} = k[P_{x-1}(t) - P_x(t)].$$

Thus, since

$$\lim_{\Delta t \to 0} \frac{P_x(t + \Delta t) - P_x(t)}{\Delta t} = P'_x(t),$$

we have

$$P'_x(t) = k[P_{x-1}(t) - P_x(t)]. \tag{4}$$

Consequently,

$$P'_0(t) = -kP_0(t),$$

since

$$P_{-1}(t) = 0.$$

Hence

$$P_0(t) = e^{-kt}. \tag{5}$$

Letting $x = 1$, in (4) we have

$$P'_1(t) = k[e^{-kt} - P_1(t)]$$

and solving, we get, noting that $P_1(0) = 0$,

$$P_1(t) = kte^{-kt}. \tag{6}$$

Continuing in this manner, we obtain

$$P_x(t) = \frac{(kt)^x e^{-kt}}{x!}. \tag{7}$$

Letting $\mu = kt$ in (7), we have

$$P_x(t) = \frac{\mu^x e^{-\mu}}{x!}, \qquad x = 0,1,\ldots,$$

which is the Poisson density function.

In the above motivation we have described a general physical situation which leads to the Poisson distribution. A specific example is the distribution of the number of failures, x, of an equipment within a time T assuming independent failures and a constant failure rate. Another such example is the distribution of the number of calls, x, arriving at a switchboard within a time T assuming independent calls and a constant arrival rate.

Example 1. Suppose a certain radar system has an MTBF (mean time between failures) of 200 hours and a constant failure rate β. Assuming failures occur independently and ignoring repair time, what is the probability of no failures in a 400-hour period?

Solution: The problem satisfies the assumptions used in deriving the Poisson distribution, and hence if x = number of failures in 400 hours, we have $\beta = \frac{1}{200}$ and

$$f_X(x) = \frac{\mu^x e^{-\mu}}{x!}, \qquad x = 0,1,2,\ldots,$$

where

$$\mu = 400\beta = 2.$$

Thus

$$f_x(0) = e^{-2} \cong .135.$$

Second Motivation We will show that the Poisson distribution is a limiting case of the binomial distribution:

$$\lim_{\substack{n\to\infty\\ p\to 0\\ np\to\mu}} \binom{n}{x} p^x q^{n-x} = \frac{e^{-\mu}\mu^x}{x!}. \tag{8}$$

The notation on the left side of (8) means that the limit as $n \to \infty$ and $p \to 0$ is to be taken along a path such that $np \to \mu$.

The proof is simple.

$$\binom{n}{x} p^x q^{n-x} = \frac{n(n-1)\cdots(n-x+1)}{x!\,n^x q^x}(np)^x q^n$$
$$= \frac{1\cdot\left(1-\frac{1}{n}\right)\cdots\left(1-\frac{x-1}{n}\right)(np)^x(1-p)^n}{x!\,q^x}.$$

But

$$\lim_{\substack{n\to\infty\\ p\to 0}} \frac{1\cdot\left(1-\frac{1}{n}\right)\cdots\left(1-\frac{x-1}{n}\right)}{x!\,q^x}(np)^x = \frac{\mu^x}{x!},$$

since

$$q \to 1, \qquad 1\cdot\left(1-\frac{1}{n}\right)\cdots\left(1-\frac{x-1}{n}\right) \to 1, \qquad \text{and} \qquad np \to \mu.$$

Now

$$(1-p)^n = [(1-p)^{-1/p}]^{-np}.$$

But,

$$\lim_{p\to 0}(1-p)^{-1/p} = e.$$

Thus

$$\lim_{\substack{n\to\infty\\ p\to 0\\ np\to\mu}} \binom{n}{x} p^x q^{n-x} = \lim_{n\to\infty}\left\{\frac{1\cdot\left(1-\frac{1}{n}\right)\cdots\left(1-\frac{x-1}{n}\right)}{x!\,q^x}\mu^x\right\}$$
$$\lim_{p\to 0}\left\{[(1-p)^{-1/p}]^{-\mu}\right\}$$
$$= \frac{\mu^x}{x!}e^{-\mu},$$

which is the desired result. Note that (8) tells us that when p

is small and n is large, we can approximate the binomial density function with the Poisson density function. A rule of thumb that is sometimes advocated for quantitizing the words "small" and "large" in the previous sentence is $p \leq .05$ and $n \geq 100$.

Example 2. If a box contains 200 items and from experience it is known that the probability that any one of these items is defective is .02, what is the probability the box will contain at least one and at most five defective items?

Solution: If x is the number of defectives, we have[1]

$$P[1 \leq X \leq 5] = \sum_{x=1}^{5} \binom{200}{x} (.02)^x (.98)^{200-x} \approx .78. \tag{9}$$

Equation (9) is a lengthy calculation to say the least. However, if we use the Poisson approximation we have

$$P[1 \leq X \leq 5] \cong \sum_{x=1}^{5} \frac{e^{-4} 4^x}{x!} = 4e^{-4}\left(1 + \frac{4}{2} + \frac{4^2}{3!} + \frac{4^3}{4!} + \frac{4^4}{5!}\right)$$
$$\cong .77.$$

4.7 THE MULTINOMIAL DISTRIBUTION

The multinomial distribution is a natural extension of the binomial distribution. Recall that the binomial distribution represents a physical process of repeated independent trials, where each trial results in an outcome S or F and the probabilities of S and F remain fixed at p and $q = 1 - p$.

Now consider a sequence of n independent trials with exactly k possible outcomes on each trial. Let the probabilities of these outcomes be denoted by $p_1, p_2, \ldots, p_k$, where $\sum_{i=1}^{k} p_i = 1$. Let $X_1, X_2, \ldots, X_k$ be random variables defined by the number of times the outcomes associated with $p_1, p_2, \ldots, p_k$ occur, re-

[1] As can be seen from this example, $P[X \leq k]$ can be difficult to calculate in the case of a binomial distribution. Tables of values of $\sum_{r=k}^{n} \binom{n}{k} p^k q^{n-k}$ have been calculated. Brief versions of such tables are included in many elementary statistics books [see, for example, Meyer (Ref. 15)].

spectively. The joint density of the random variables $X_1, \ldots, X_{k-1}$ is then given by

$$f_{X_1, \ldots, X_{k-1}}(x_1, \ldots, x_{k-1}) = \begin{cases} \dfrac{n!}{x_1! x_2! \cdots x_k!} p_1^{x_1} p_2^{x_2} \cdots p_k^{x_k}, \\ x_i = 0, 1, \ldots, n, \sum_{i=1}^{k} x_i = n, \\ 0, \qquad \text{elsewhere.} \end{cases} \tag{1}$$

The distribution defined by (1) is called the multinomial distribution. We have indicated it to be a function only of $x_1, \ldots, x_{k-1}$ since when these values are known, x_k is known by the relation $x_k = n - \sum_{i=1}^{k-1} x_i$.

It can be shown that the marginal density of X_i in (1) is binomial with mean p_i by summing (1) over the remaining x's. Thus it follows that

$$E[X_i] = np_i$$

and

$$\text{Var } X_i = np_i(1 - p_i).$$

Example 1. A rod of a specific length is manufactured. Assume that the actual length is best thought of as a value of a random variable uniformly distributed over the interval [4,5]. Now suppose we are interested only in determining the probability of occurrence of the three events

$$E_1 = \{X \leq 4.45\}, \quad E_2 = \{4.45 \leq X \leq 4.50\}, \quad E_3 = \{X > 4.50\}.$$

If 20 such rods are manufactured, what is the probability that exactly 12 will be of length less than 4.45 units, and exactly 15 will exceed 4.50 units?

Solution. We first note that

$$P[E_1] = \int_4^{4.45} dx = .45,$$

Table 4.1

Density Function	Functional Description	Mean	Variance	Moment-Generating Function
Point binomial	$f_X(x) = \begin{cases} p^x q^{1-x}, & x = 0,1, \\ 0, & \text{elsewhere} \end{cases}$ $p + q = 1$	p	pq	$pe^t + q$
Binomial	$f_X(x) = \begin{cases} \binom{n}{x} p^x q^{n-x}, & x = 0,1,\ldots,n, \\ 0, & \text{elsewhere} \end{cases}$ $p + q = 1$	np	npq	$(pe^t + q)^n$
Negative binomial	$f_X(x) = \begin{cases} \binom{x+r-1}{r-1} p^r q^x, & x = 0,1,\ldots, \\ 0, & \text{elsewhere} \end{cases}$	$\frac{rq}{p}$	$\frac{rq}{p^2}$	$\left(\frac{p}{1 - e^t q}\right)^r$
Hypergeometric	$f_X(x) = \begin{cases} \frac{\binom{m}{x}\binom{k}{n-x}}{\binom{m+k}{n}}, & x = 0,\ldots,n, \\ 0, & \text{elsewhere} \end{cases}$	np	$\frac{npq(m+k-n)}{m+k-1}$	$\sum_{x=0}^{n} e^{tx} \frac{\binom{m}{x}\binom{k}{n-x}}{\binom{m+k}{n}}$
Poisson	$f_X(x) = \begin{cases} \frac{e^{-\mu}\mu^x}{!}, & x = 0,1,\ldots, \\ 0, & \text{elsewhere} \end{cases}$	μ	μ	$e^{\mu(e^t - 1)}$

$$P[E_2] = \int_{4.45}^{4.50} dx = .05,$$

$$P[E_3] = \int_{4.50}^{5} dx = .50.$$

Thus, by (1) the desired probability p is

$$p = \frac{20!}{12!3!15!}(.45)^{12}(.05)^3(.5)^{15}.$$

4.8 THE NORMAL DISTRIBUTION

In the previous sections of this chapter we limited our discussions of distributions to those generated by discrete density functions. We will now turn our attention to continuous distributions.

The most important and most extensively studied distribution is the normal distribution (see Figures 4.1 and 4.2). It is defined as follows. Let X be a random variable with pdf f_X defined by

$$f_X(x) = \frac{1}{b\sqrt{2\pi}} \exp\left[-\frac{1}{2}\left(\frac{x-a}{b}\right)^2\right], \qquad -\infty < x < \infty. \tag{1}$$

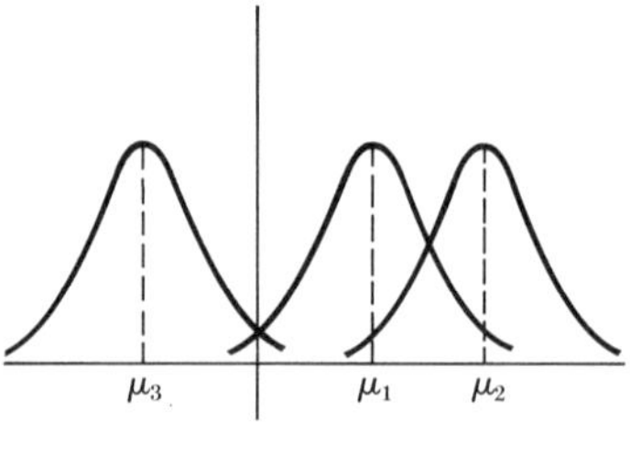

Figure 4.1 Normal Curves with Different Means but the Same Variance

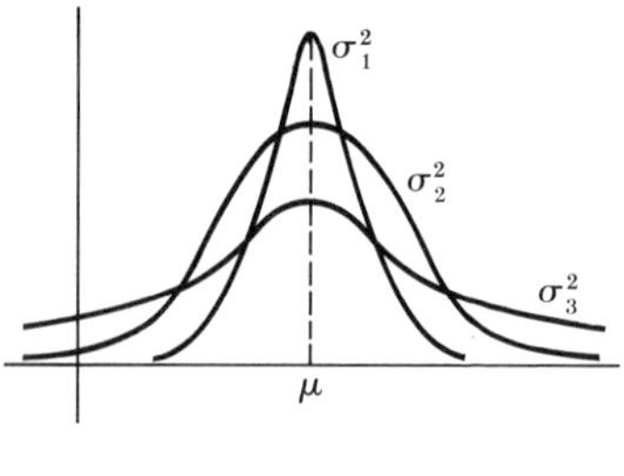

Figure 4.2 Normal Curves with the Same Mean and Variances $\sigma_1^2 < \sigma_2^2 < \sigma_3^2$

Then we say X has a normal distribution with parameters a and b. Recall that in Example 4, Section 3.3, we showed that $a = \mu_X$ and $b = \sigma_X$. Thus we write

$$f_X(x) = \frac{1}{\sigma_X\sqrt{2\pi}} \exp\left[-\frac{1}{2}\left(\frac{x - \mu_X}{\sigma_X}\right)^2\right], \qquad -\infty < x < \infty, \tag{2}$$

and we say $X \sim N(\mu_X, \sigma_X^2)$, which is read "$X$ is distributed normal with mean μ_X and variance σ_X^2."

We have already established (see Example 2, Section 3.5) the moment-generating function when $X \sim N(\mu_X, \sigma_X^2)$. We list it again here.

$$M_X(t) = \exp(\mu_X t + \sigma_x^2 t^2/2) \tag{3}$$

This result can be used to prove the following important result, which we list as a theorem.

Theorem 1 Let $X_1, \ldots, X_n$ be a random sample. Then $\bar{X}$ has a normal distribution with mean μ_X and variance σ_X^2/n; that is,

$$\bar{X} \sim N(\mu_X, \sigma_X^2/n). \tag{4}$$

A similar theorem of equal importance which can also be shown from (3) is the following.

Theorem 2 Let $X \sim N(\mu, \sigma^2)$ and

$$Y = \frac{X - \mu}{\sigma}. \tag{5}$$

Then

$$Y \sim N(0, 1).$$

Proof:

$$M_Y(t) = M_{(X-\mu)/\sigma}(t) = e^{-\mu t/\sigma} M_X\left(\frac{t}{\sigma}\right) = e^{(1/2)t^2}. \tag{6}$$

But (6) is the moment-generating function of a $N(0,1)$ random variable and the theorem follows.

Theorem 2 is useful in tabulating the cumulative normal distribution, since due to this result it is only necessary to tabulate the $N(0,1)$ distribution or the so-called standard normal distribution. Table I in the Appendix is such a table, that is, a table of values of $F_X(x)$ where

$$F_X(x) = \frac{1}{\sqrt{2\pi}} \int_{-\infty}^{x} e^{-(1/2)y^2}\, dy. \tag{7}$$

To demonstrate its use, consider the following two examples.

Example 1. If $X \sim N(0,1)$, find the $P[X < 1]$, $P[X < -1]$, and $P[-1 < X < 1]$.

Solution:

$$P[X < 1] = F_X(1) = .8413$$

by Table I.

Second, note that

$$P[X < -1] = P[X > 1].$$

To see this latter statement note Figure 4.3. Now $P[X < -1]$ is simply the area to the left of -1, while $P[X > 1]$ is the area to the right of 1. By symmetry it is clear that $P[X < -1] = P[X > 1]$.

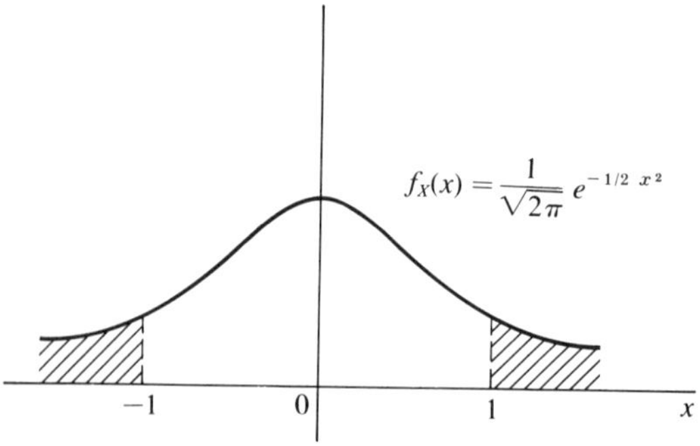

Figure 4.3 Standardized Normal

But then

$$P[X < -1] = P[X > 1] = 1 - P[X \leq 1] = 1 - F_X(1)$$
$$= .1587.$$

Finally,

$$P[-1 < X < 1] = F_X(1) - F_X(-1)$$
$$= .6826.$$

Example 2. If $X \sim N(5,4)$ find $P[X < 7]$, $P[X < 3]$, and $P[3 < X < 7]$.

Solution: Note that in this case $\mu = 5$, $\sigma = 2$; that is, $\sigma^2 = 4$. Then remembering that $(X - 5)/2 \sim N(0,1)$, we have

$$P[X < 7] = P\left[\frac{X-5}{2} < \frac{7-5}{2}\right]$$
$$= P\left[\frac{X-5}{2} < 1\right]$$
$$= .8413$$

and

$$P[X < 3] = P\left[\frac{X-5}{2} < \frac{3-5}{2}\right]$$
$$= P\left[\frac{X-5}{2} < -1\right]$$
$$= .1587,$$

so that

$$P[3 < X < 7] = .8413 - .1587 = .6826.$$

Example 2 illustrates two important facts. The first is that if $X \sim N(\mu,\sigma^2)$, we can always find $F_X(x)$ from a $N(0,1)$ table. This follows since

$$F_X(x) = P[X \leq x] = P\left[\frac{X-\mu}{\sigma} \leq \frac{x-\mu}{\sigma}\right]$$

for any x. But the random variable $(X - \mu)/\sigma \sim N(0,1)$ and hence the probability on the right can be obtained from Table I.

The second fact is that the statement

$$F_X(x) = \alpha,$$

when $X \sim N(0,1)$, implies that

$$F_Y(\mu + x\sigma) = \alpha,$$

if

$$Y \sim N(\mu,\sigma^2);$$

that is, when $Y \sim N(\mu,\sigma^2)$ Table I gives the area under $f_Y(x)$ which lies to the left of $\mu + x\sigma$.

By Table I, for example, $P[X \leq 2] = .9772$, and hence, in Figure 4.4, 97.72 percent of the area lies to the left of $\mu + 2\sigma$.

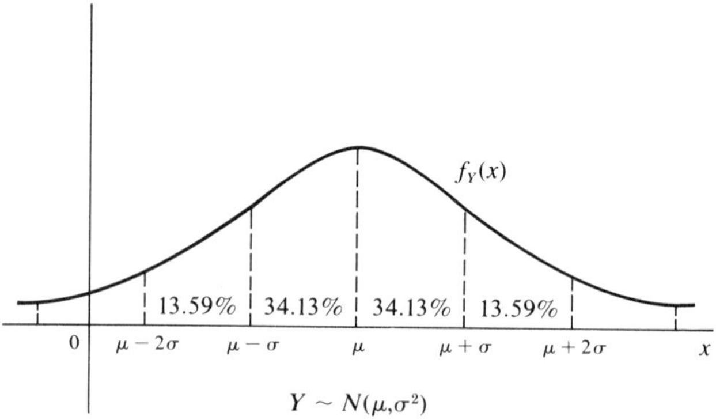

Figure 4.4 Normal Density Function

Similarly, 97.72 percent of the area lies to the right of $\mu - 2\sigma$ and hence $P[\mu - 2\sigma \leq X \leq \mu + 2\alpha] = .9544$.

As we mentioned while introducing the normal distribution, it is the most important and thoroughly studied of all distributions. We now state a theorem, which justifies this intensive study. We remark that the theorem is one of the most remarkable and beautiful results of mathematics.

Theorem 3 (The Central Limit Theorem). Let $X_1, \ldots, X_n$ be a random sample from a distribution with finite mean and variance. Let

$$\bar{X} = \frac{1}{n} \sum_{i=1}^{n} X_i$$

and

$$Y_n = \frac{\bar{X} - \mu_X}{\sigma_X} \sqrt{n}.$$

Then f_{Y_n}, the pdf of Y_n, approaches a $N(0,1)$ distribution as $n \to \infty$; that is,

$$f_{Y_n}(y) \to \frac{1}{\sqrt{2\pi}} e^{-(1/2)y^2}, \qquad -\infty < y < \infty.$$

Note that no assumption other than finite mean and variance has been made in regard to the distribution of X. Thus, for most practical situations the distribution of $\bar{X}_n$ is essentially normal for n sufficiently large. Of course, the question of how fast does this convergence take place is a pertinent one. The rapidity of the convergence is a function of the distribution of X. In many cases, however, $n \geq 10$ is sufficient. In fact, sampling experiments have shown that if $n > 50$, the density of X has little effect on the distribution of $\bar{X}_n$. If X has a point binomial distribution, Theorem 2 tells us that the distribution of the quantity

$$\frac{\bar{X} - p}{\sqrt{pq}} \sqrt{n} \to N(0, 1) \qquad \text{as } n \to \infty. \tag{8}$$

But as shown in Section 4.2, the binomial distribution with mean np and variance npq results as the distribution of the sum, $n\bar{X}$, generated by a random sample from a point binomial distribution with mean p and variance pq. Now if we multiply the numerator and denominator of (8) by n we have the distribution of

$$\frac{n\bar{X} - np}{n\sqrt{pq/n}} = \frac{n\bar{X} - np}{\sqrt{npq}} \to N(0,1) \qquad \text{as } n \to \infty. \tag{9}$$

Since the reader may not see the full impact of (9), we now restate it as a corollary to Theorem 2.

Corollary 1 If

$$f_X(x) = \begin{cases} \binom{n}{x} p^x q^{n-x}, & x = 0,1,\ldots,n, \\ 0, & \text{elsewhere,} \end{cases}$$

and

$$Y_n = \frac{X - np}{\sqrt{npq}}, \tag{10}$$

then the distribution of Y_n approaches $N(0,1)$ as $n \to \infty$; i.e.,

$$\lim_{n \to \infty} f_{Y_n}(y) = \frac{1}{\sqrt{2\pi}} e^{-(1/2)y^2}, \qquad -\infty < y < \infty.$$

We will now demonstrate an application of this theorem by means of the following example.

Example 3. Let

$$f_X(x) = \binom{100}{x}(.2)^x(.8)^{100-x}, \qquad x = 0,1,\ldots,100.$$

Find the $P[10 < X \leq 20]$.

Solution: Clearly

$$P[10 < X \leq 20] = \sum_{x=11}^{20} \binom{100}{x}(.2)^x(.8)^{100-x}.$$

To approximate the above probability, we use Corollary 1. By that corollary we have

$$\begin{aligned} P[10 < X \leq 20] &= P\left[\frac{10 - np}{\sqrt{npq}} < \frac{X - np}{\sqrt{npq}} \leq \frac{20 - np}{\sqrt{npq}}\right] \\ &\approx \frac{1}{\sqrt{2\pi}} \int_{(10-np)/\sqrt{npq}}^{(20-np)/\sqrt{npq}} e^{-(1/2)t^2}\, dt. \end{aligned}$$

In this example $np = 100(.2) = 20$ and $\sqrt{npq} = \sqrt{16} = 4$.

Thus

$$P[10 < X \leq 20] \approx \frac{1}{\sqrt{2\pi}} \int_{-2.5}^{0} e^{-(1/2)t^2}\, dt = .494,$$

by Table I.

This approximation can be improved if we reason as follows. Since we are approximating a discrete distribution with a continuous distribution, it seems logical to regard $10 < X \leq 20$ as $10.5 \leq X \leq 20.5$. We then obtain

$$P[10 < X \leq 20] \approx \frac{1}{\sqrt{2\pi}} \int_{-2.38}^{.13} e^{-(1/2)t^2}\, dt = .503,$$

where the limits have been determined by the relations

$$\frac{20.5 - np}{\sqrt{npq}} \approx .13$$

and

$$\frac{10.5 - np}{\sqrt{npq}} \approx -2.38.$$

An additional corollary, which is a consequence of Corollary 1 and which can be obtained by dividing the numerator and denominator in equation (10) by n, is as follows.

Corollary 2 If x is the number of successes in n binomial trials, then the distribution of the proportion of successes, x/n, is approximately $N(p, pq/n)$ when n is sufficiently large.

Example 4. An engineer involved in the mass production of a large number of items desires to construct a quality-control chart to automatically determine when system malfunctions are present. Suppose that when the production line is operating satisfactorily, 2.5 percent (on the average) of a day's production is defective. Determine a procedure so that 95 percent of the time one can detect when the proportion of defectives has increased, and hence the system may need maintenance.

The difficulty in asserting when the production line is malfunctioning lies in the fact that for any given n (that is, number of daily manufactured items), the proportion of defectives in the day's output is random in nature. That is, although the system produces only 2.5 percent defectives on the average, on any particular day the number of defectives may be more or less than this, even though the production line is functioning quite properly. Of course, if the entire day's production is defective, it is quite obvious that something is wrong. Normally, however, the reduction in quality is not this dramatic, and we desire a procedure which will alert us to the fact that the quality of the day's production has slipped below what can be attributed to chance. To solve the problem, we consider the finished products as outcomes of a set of binomial trials. Then the proportion of defectives x/n in a day's output of n items is approximately normally distributed with mean p and variance pq/n.

Thus we see that if $p = .025$, x/n should fall within the interval $[.025 - 1.96\sigma, .025 + 1.96\sigma]$ 95 percent of the time, since for large n, x/n is approximately distributed $N(p, pq/n)$, and therefore

$$P\left[\mu_X - 1.96\sigma \leq \frac{x}{n} \leq \mu_X + 1.96\sigma\right] \cong .95.$$

But

$$\sigma = \frac{pq}{n} = \frac{(.025)(.975)}{n} = \frac{.024}{n}.$$

It follows then that 95 percent of the time x/n should fall within the interval $[.025 - (.047/n), .025 + (.047/n)]$. We can now construct the chart shown in Figure 4.5.

Charts such as Figure 4.5 are called quality-control charts. Note that by simply plotting the proportion of defectives against days we have a time record of the quality performance. Further, if x/n exceeds the line $.025 + (.047/n)$, then we have what might be construed as sufficient evidence to claim that the production line needs some corrective maintainance. Also, if x/n falls below the line $.025 - (.047/n)$, we could assert that the quality of production has improved.

In the above we have assumed that the number of items produced daily was fixed at n. This was assumed only for convenience and no real additional difficulty evolves if n varies from day to day.

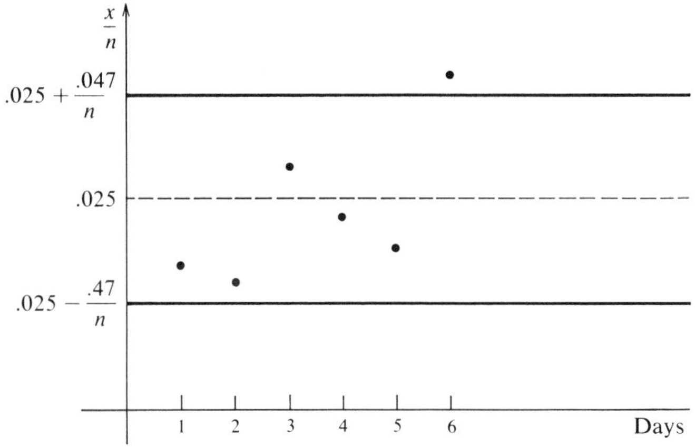

Figure 4.5 Control Chart

Example 5. Suppose the army issued 30,000 new pairs of boots to new draftees. If experience indicates that 25 percent of these pairs of boots will need replacing in six months, how many pairs of boots should they have on hand for this purpose in order to be 99 percent sure that a sufficient number of boots will be available?

Solution: We can consider the replacement or nonreplacement of a pair of boots as the possible outcome of a sequence of binomial trials. If X is a random variable whose possible values are the number of pairs of boots which need replacing, then X has a binomial distribution with mean np and variance npq, where $p = \frac{1}{4}$, $q = \frac{3}{4}$, and $n = 30{,}000$. However, for large n, X is approximately normal with mean np and variance npq. But from Table I, 99 percent of the area under a normal curve lies to the left of $\mu + 2.326\sigma$. Hence 99 percent of the time the number of pairs of boots which need replacing will not exceed $30{,}000\,(\frac{1}{4}) + 2.326\sqrt{(30{,}000)(\frac{3}{16})}$ or 7675.

4.9 The Chi-Square Distribution

In Section 2.8 we discussed the fact that in many cases one's interest settles on the distribution of the sum of the squares of a collection of random variables. This is particularly true in the case of a collection of $N(0, 1)$ random variables. The distribution which arises in this case is called the *chi-square distribution*. It is defined as follows. Let X be a random variable such that

$$f_X(x) = \begin{cases} \dfrac{x^{(n/2)-1}e^{-(1/2)x}}{\Gamma\left(\dfrac{n}{2}\right)2^{n/2}}, & x \geq 0 \\ 0, & \text{elsewhere.} \end{cases}$$

Then X is said to have a chi-square distribution[1] with n degrees of freedom or $X \sim \chi^2(n)$.

The importance of the chi-square distribution arises from the following closely associated theorems, which are proved in Mood and Graybill (Ref. 16).

Theorem 1 Let $X_1, \ldots, X_n$ be a random sample from a $N(0, 1)$ distribution. Then the random variable

$$Y = \sum_{i=1}^{n} X_i^2 \sim \chi^2(n). \tag{1}$$

Corollary 1 If $X_1, \ldots, X_n$ is a random sample from a $N(\mu, \sigma^2)$ distribution, then

$$Y = \sum_{i=1}^{n} \left(\frac{X_i - \mu}{\sigma}\right)^2 \sim \chi^2(n). \tag{2}$$

Theorem 2 Let $X_1, \ldots, X_n$ be a random sample from a $N(\mu, \sigma^2)$ distribution; then $\overline{X}$ and S^2 are independent random variables and $nS^2/\sigma^2 \sim \chi^2(n - 1)$. That is,

$$\frac{nS^2}{\sigma^2} = \frac{1}{\sigma^2}\sum_{i=1}^{n}(X_i - \overline{X})^2 \tag{3}$$

is distributed chi square with $n - 1$ degrees of freedom.

---

[1]The quantity Γ is called the gamma function and is defined by

$$\Gamma\left(\frac{n}{2}\right) = \int_0^\infty e^{-t}t^{(n/2)-1}\,dt.$$

The function is an extension of the factorial function and it is easy to show that $\Gamma(k) = (k - 1)!$ for any integer k. The gamma function has been extensively studied and tabulated. See Rainville (Ref. 20).

It should be noted that in the above theorem we have stated the fact that for normal populations the sample mean and sample variance are independently distributed. It is an interesting and important fact that no other distribution has this property.

The moment-generating function of a $\chi^2(n)$ variable can be obtained by means of Theorem 1 as follows.

First, if $X \sim N(0, 1)$, then

$$\begin{aligned} M_{X^2}(t) &= E[e^{X^2 t}] = \frac{1}{\sqrt{2\pi}} \int_{-\infty}^{\infty} e^{x^2 t} e^{-(1/2)x^2}\, dx \\ &= \frac{1}{\sqrt{2\pi}} \int_{-\infty}^{\infty} e^{-(1/2)(1-2t)x^2}\, dx \\ &= (1 - 2t)^{-1/2}, \qquad t < \tfrac{1}{2}. \end{aligned} \tag{4}$$

But, by Theorem 1, $Y = \sum_{i=1}^{n} X_i^2 \sim \chi^2(n)$, and, by (4),

$$M_Y(t) = [M_{X_i^2}(t)]^n = (1 - 2t)^{-n/2}. \tag{5}$$

Thus equation (5) gives the moment-generating function for any $\chi^2(n)$ random variable.

From (5) it is trivial to prove the sum of independent chi-square variables is again a chi-square variable.

Theorem 3 Let $X_1, X_2, \ldots, X_n$ be independent χ^2 random variables with degrees of freedom $m_1, m_2, \ldots, m_n$, respectively. Then

$$Y = \sum_{i=1}^{n} X_i \sim \chi^2\left(\sum_{i=1}^{n} m_i\right);$$

that is, Y has a chi-square distribution with $m_1 + m_2 + \cdots + m_n$ degrees of freedom (see Figure 4.6). Since the proof is so simple we will include it.

Proof:

$$\begin{aligned} M_Y(t) &= M_{X_1}(t) M_{X_2}(t) \cdots M_{X_n}(t) \\ &= (1 - 2t)^{-m_1/2} \cdots (1 - 2t)^{-m_n/2} \\ &= (1 - 2t)^{-(1/2)\sum_{i=1}^{n} m_i}. \end{aligned} \tag{6}$$

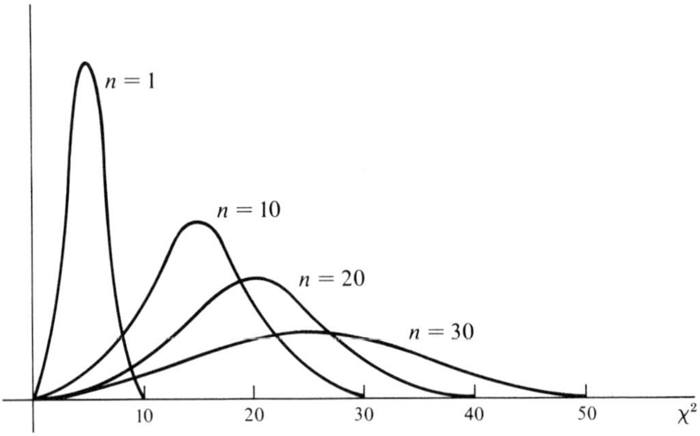

Figure 4.6 Chi-Square Distribution for Different Values of n

But (6) is the moment-generating function of a chi-square variable with $\sum_{i=1}^{n} m_i$ degrees of freedom, and Theorem 3 follows.

We can also use (5) to obtain the mean and variance of Y. Thus

$$\mu_Y = \frac{d}{dt}(1 - 2t)^{-n/2}\Big|_{t=0} = n$$

and

$$\sigma_Y^2 = \frac{d^2}{dt^2}(1 - 2t)^{-n/2}\Big|_{t=0} - n^2 = 2n.$$

Table II in the Appendix gives the area from 0 to x under the chi-square curve for various values of x and n.

4.10 The Cauchy Distribution

If the random variable X has the pdf

$$f_X(x) = \frac{\alpha}{\pi} \frac{1}{\alpha^2 + (x - \mu)^2}, \qquad -\infty < x < \infty,$$

then we say that X has a Cauchy distribution. The Cauchy distribution differs somewhat from the previous distributions discussed in that its mean only exists as a principal value and no higher moments exist (see Figure 4.7). The distribution arises as the distribution of the ratio of two independent normal variables with zero means. That is, if $X \sim N(0, \sigma_X^2)$, $Y \sim N(0, \sigma_Y^2)$ and X and Y are independent, then

$$Z = \frac{X}{Y} \sim \text{Cauchy with } \mu = 0 \text{ and } \alpha = \frac{\sigma_X}{\sigma_Y}.$$

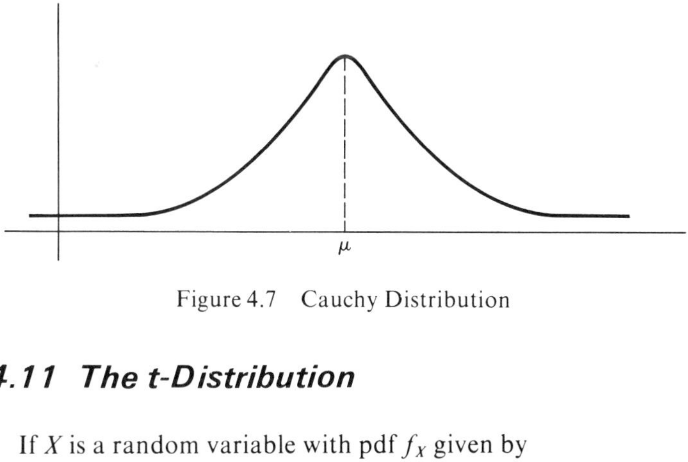

Figure 4.7 Cauchy Distribution

4.11 The t-Distribution

If X is a random variable with pdf f_X given by

$$f_X(x) = \frac{\Gamma\left(\frac{k+1}{2}\right)}{\Gamma\left(\frac{k}{2}\right)\sqrt{\pi k}}\left(1 + \frac{t^2}{k}\right)^{-(k+1)/2}, \qquad -\infty < t < \infty, \qquad (1)$$

then X is said to have a t-distribution with k degrees of freedom. The distribution derives its importance from the following theorem.

Theorem 1 Let X and Y be independent random variables such that

$$X \sim N(0, 1) \qquad \text{and} \qquad Y \sim \chi^2(k).$$

If

$$Z = \frac{X}{\sqrt{Y}} \sqrt{k}, \tag{2}$$

then Z has a t-distribution with k degrees of freedom. We denote this by $Z \sim t(k)$. By making use of this theorem it is possible to put a confidence interval on the mean of a normal distribution without knowing the variance. We will discuss this in more detail in Chapter V. For k large (say $k > 100$) the t-distribution is essentially the same as a $N(0, 1)$ distribution. In fact,

$$\lim_{k \to \infty} \frac{\Gamma\left(\frac{k+1}{2}\right)}{\Gamma\left(\frac{k}{2}\right)\sqrt{\pi k}} \left(1 + \frac{t^2}{k}\right)^{-(k+1)/2} = \frac{1}{\sqrt{2\pi}} e^{-(1/2)t^2}$$

The approach to normality is shown in Figure 4.8.

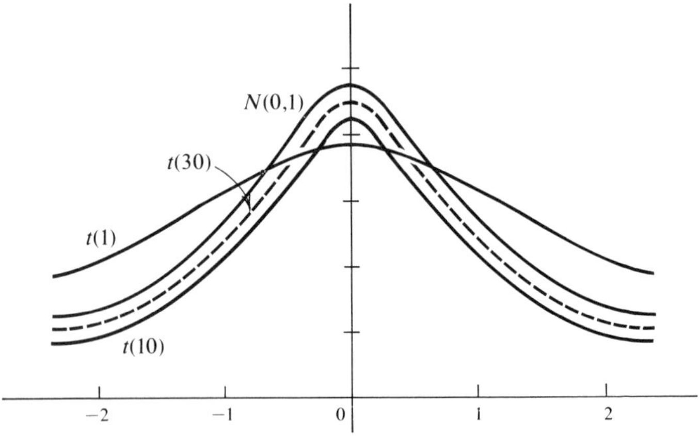

Figure 4.8 t-Distribution Density Function

4.12 THE F-DISTRIBUTION

A random variable X is said to have an F-distribution with m and n degrees of freedom if

$$f_X(x) = \begin{cases} \dfrac{\Gamma\left(\dfrac{m+n}{2}\right)}{\Gamma\left(\dfrac{m}{2}\right)\Gamma\left(\dfrac{n}{2}\right)}\left(\dfrac{m}{n}\right)^{m/2} \dfrac{x^{(m-2)/2}}{\left(1 + \dfrac{mx}{n}\right)^{(m+n)/2}}, & x > 0, \\ 0, & x \leq 0. \end{cases} \tag{1}$$

We will discuss this distribution further in Chapter V. It derives its importance from the following theorem, which arises in the problem of estimating the ratio of two variances of two different normal distributions.

Theorem 1 Let U and V be independent chi-square random variables with m and n degrees of freedom, respectively. If

$$X = \frac{U/m}{V/n}, \tag{2}$$

then X has an F-distribution with m and n degrees of freedom; that is, $X \sim F(m, n)$.

Corollary If $X \sim F(m, n)$, then $X^{-1} \sim F(n, m)$.

The Distribution of Reliability In Sections 4.14, 4.15, 4.16, and 4.17 we will discuss four distributions which have found considerable application in the field of reliability engineering. We would like to hasten to add that this is not to be interpreted as implying that these distributions have no application in other areas. Indeed, most of them have found use in a diversity of fields ranging from economics to chemistry. However, they do play an important role in reliability engineering and this particular subject affords us a uniform and natural way of introducing each of the distributions as the appropriate probability model for a given physical process. To accomplish this it is necessary to discuss briefly the reliability and hazard functions.

4.13 Reliability and Hazard Functions

In describing the failure properties of a tangible object it is clear that the length of the life[1] of such objects cannot be

[1] By length of life we are referring to the length of the interval for which an object functions at some defined level of acceptability.

described by deterministic methods. For example, two "identical" fuses, steel beams, motors, human beings, radar systems, etc., subjected to the same stresses will almost always expire at different and unpredictable times. It is necessary, therefore, to consider the time to failure or length of life T of an object as a continuous random variable. In connection with this formulation of the problem we make the following definitions, in which physical objects are termed "systems."

Definition 1 Let T be a random variable defined by the time to failure (or length of life) of a system. Then we define the reliability function $R(t)$ by

$$R(t) = P[T > t] = 1 - F_T(t). \tag{1}$$

Thus we see that $R(t)$ is the probability that the system will survive from time zero to at least time t, or equivalently that it will not fail before time t.

Definition 2 Let f_T be the pdf of T. Then the hazard function $H(t)$ associated with the random variable T is defined by

$$H(t) = \frac{f_T(t)}{1 - F_T(t)} = \frac{f_T(t)}{R(t)}, \tag{2}$$

when $R(t) \neq 0$.

The function defined in equation (2) is also called the instantaneous failure rate. The reason for this latter name is seen by noting

$$\begin{aligned} P[T \leq t + \Delta t \mid T \geq t] &= \frac{P[t \leq T \leq t + \Delta t]}{P[T \geq t]} \\ &= \frac{P[t \leq T \leq t + \Delta t]}{1 - F_T(t)}, \end{aligned}$$

where the left side is the probability that the system will not survive to time $t + \Delta t$ given that it has survived to time t. Thus

$$P[T \leq t + \Delta t \mid T \geq t] = \frac{\int_t^{t+\Delta t} f_T(x)\,dx}{1 - F_T(t)}$$

$$= \frac{\Delta t f_T(\xi)}{1 - F_T(t)}$$

for some ξ such that $t \leq \xi \leq t + \Delta t$. Hence, if f_T is continuous at t we see by (2) that for Δt small,

$$P[T \leq t + \Delta t \mid T \geq t] \approx \Delta t H(t).$$

We have therefore, established that at any time t the probability that a functioning system will fail in the next Δt units is approximately $H(t)\, \Delta t$, and therefore justified use of the phrase instantaneous failure rate. Of course, there is a larger interpretation of what we have just shown. More generally we have shown that the hazard rate is effectively the instantaneous rate at which any time-ordered events occur.

The importance of the hazard function is primarily due to the fact that it uniquely determines the density function. That is, by determining the instantaneous failure rate or hazard function for a system we determine uniquely the pdf of time to failure. This is seen by means of the following theorem.

Theorem 1 Let T be a continuous random variable such that $F_T(t) = 0$ if $T \leq 0$ with hazard function $H(t)$. Then the pdf f_T is given by

$$f_T(t) = \begin{cases} H(t) \exp\left[-\int_0^t H(x)\,dx\right], & t \geq 0, \\ 0, & t < 0. \end{cases} \tag{3}$$

The theorem is easily proved by noting from (2) that

$$H(t) = \frac{f_T(t)}{R(t)} = \frac{-R'(t)}{R(t)} \tag{4}$$

and thus

$$\ln R(t) = -\int_0^t H(x)\,dx$$

or

$$R(t) = \exp\left[-\int_0^t H(x)\,dx\right],$$

so that, by (4),

$$f_T(t) = R(t)H(t) = H(t) \exp\left[-\int_0^t H(x)\, dx \right].$$

Hazard functions take a variety of shapes and forms. One shape that seems rather ideal is the "bathtub" shape, shown in Figure 4.9. The physical interpretation of the figure is quite obvious. That is, initially the failure rate is quite large, but it decreases to t, where it levels out and remains essentially constant until time t_2, where it increases from there on. The time from $t = 0$ to $t = t_1$ is called the burn-in period, while the time $t \geq t_2$ is called the wear-out period. By making different assumptions in regard to the hazard function it is possible to obtain all the important time-to-failure distributions.

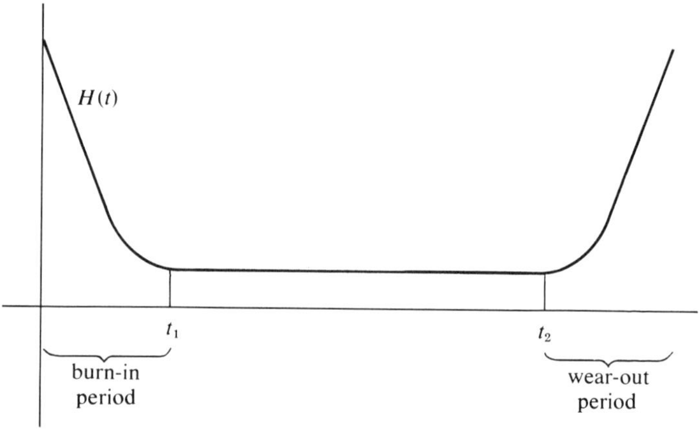

Figure 4.9 A Useful Hazard Function

4.14 THE EXPONENTIAL DISTRIBUTION

If $H(t) = \alpha$, α a constant, then

$$f_T(t) = \begin{cases} \alpha e^{-\alpha t}, & t \geq 0, \alpha > 0, \\ 0, & t < 0, \end{cases} \tag{1}$$

by (3) of Section 4.13.

The distribution defined by (1) is called the exponential distribution and is one of the most popular distributions of reliability. Its popularity is due, essentially, to three factors. First, it is easy to handle mathematically. Second, it has often been supported as a time-to-failure model through experimental data. Finally, it has some theoretical justification. The theoretical justification we speak of is from two sources. The most interesting of these mathematically is that the exponential distribution enjoys a central-limit type of theorem. That is, it can be shown that it is the limiting case for a system with a large number of in-series components, all of which have only a small probability of failing, even if the individual components do not have an exponential-failure law. The other justification is clear from Figure 4.9. From this figure we see that in a system which has a "bathtub" failure rate (such as people), the exponential distribution is a very good pdf for time to failure if the system has already been put through a burn-in period and is not operated into the wear-out period. Thus, for example, if the tubes in an electronic receiver are pretested and replaced periodically before wear-out by pretested tubes and the replacement process is continually repeated, it is reasonable to assume that the failure rates of the tubes in the receiver are constant and hence that the time to failure of each is exponentially distributed.

From (1) it is clear that the exponential distribution is completely determined by the parameter α (see Figure 4.10).

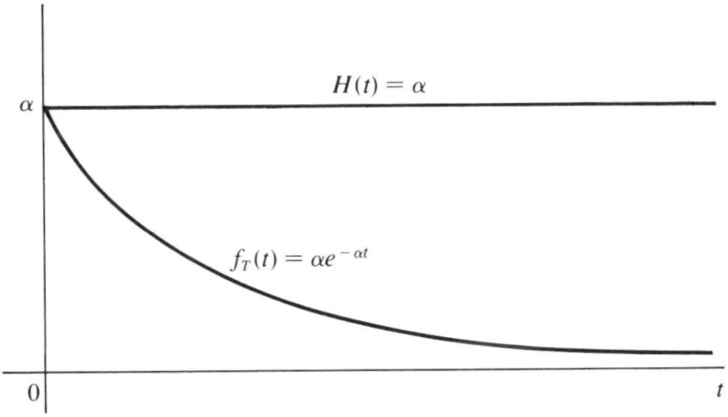

Figure 4.10 Exponential pdf and Its Hazard Function

This parameter can be related to the mean and variance of the distribution very easily by direct integration or by determining the moment-generating function. We shall do the latter.

$$M_T(\theta) = \alpha \int_0^\infty e^{\theta t} e^{-\alpha t}\, dt = \frac{\alpha}{\alpha - \theta}, \qquad \theta < \alpha. \tag{2}$$

But then

$$E[T] = \frac{d}{d\theta}\left(\frac{\alpha}{\alpha - \theta}\right)\Bigg|_{\theta=0} = \frac{1}{\alpha} \tag{3}$$

and

$$\sigma_T^2 = \frac{d^2}{d\theta^2}\left(\frac{\alpha}{\alpha - \theta}\right)\Bigg|_{\theta=0} - \frac{1}{\alpha^2} = \frac{1}{\alpha^2}. \tag{4}$$

Example 1. Find the reliability function for an electronic generator whose MTBF = 5000 hours. (MTBF, mean time between failures, is the terminology usually used in reliability for $E[T]$.) Assume that the system has been previously burned in.

Solution: Under the assumption of a burn-in being previously performed we can solve our problem for all t which do not carry us into the wear-out phase. That is possible since then we can logically assume that the failure rate is constant and hence that $T \sim \alpha e^{-\alpha t}$. Thus

$$R(t) = 1 - F_T(t) = 1 - \int_0^t \alpha e^{-\alpha x}\, dx$$
$$= e^{-\alpha t} = e^{-t/5000}.$$

Note that

$$R(5000) = e^{-1} \approx .37,$$

and hence only about 37 percent of a large number of such generators would be expected to still be functioning satisfactorily after 5000 hours of use, even though 5000 hours is the expected life of a generator.

Example 2. In the example above, how large should the MTBF be in order that $R(5000) = .5$?

Solution:

$$R(5000) = e^{-\alpha(5000)}.$$

Hence

$$e^{-\alpha(5000)} = .5$$

or

$$\alpha = -\frac{\ln 2}{5000},$$

so that

$$\text{MTBF} = \frac{5000}{-\ln 2}\text{ hr} \approx 7214\text{ hr}.$$

In Example 1 we assumed α to be known. Practically speaking this will never be the case. However, it is possible to estimate α by the method of maximum likelihood discussed in Chapter V.

We should finally comment that there is another very nice way to obtain the exponential distribution as an appropriate probability model for time to failure. Suppose we allow a system to be repaired after each failure and assume that after repair the system is as good as new (or more precisely the same as new). Then if we ignore repair times, we have a system which can have any number of failures in time t. A similar situation is obtained by assuming that we have a number of identical systems operating at once. Now, if we assume that failures occur independently and according to a Poisson distribution with mean $\mu(t) = kt$ for each t, then it can be shown that the time between failures is exponentially distributed with $\alpha = k$. That is, if the number of failures in each interval $(0, t)$ occur according to a Poisson law at the rate k, then the time to failure is an exponential random variable with mean $1/k$. This motivation, of course, admits the same generalization as the hazard function; that is, we can replace the word "failures" by "events," and the exponential distribution then results as the distribution of the time between events.

4.15 THE GAMMA DISTRIBUTION

Now consider a system in which the failure of a particular component results in a failure of the system. However, suppose the system has in reserve $r - 1$ such components, so that the instant one fails a new one takes its place. Then a system failure due to this component occurs only after r of them have failed. Now if $T_1, \ldots, T_r$ are random variables such that the values of T_j, $j = 1, 2, \ldots, r$, are the times to failure associated with the r components, then the system time to failure t is given by values of the random variable:

$$T = \sum_{i=1}^{r} T_i. \tag{1}$$

If the failures occur independently and at the same constant instantaneous failure rate α, then $T_1, T_2, \ldots, T_r$ form a random sample from an exponential distribution with mean $1/\alpha$. To find the probability model for time to failure or length of life for this system we need to determine the pdf of T in (1). The moment-generating function for the random variable T is given by

$$M_T(\theta) = M_{T_1}(\theta) M_{T_2}(\theta) \cdots M_{T_r}(\theta),$$

and, by the previous section we have

$$M_T(\theta) = \left(\frac{\alpha}{\alpha - \theta}\right)^r. \tag{2}$$

Thus, from (6) of Section 3.5 we have

$$f_T(t) = L^{-1}\left\{\left(\frac{\alpha}{s + \alpha}\right)^r\right\}, \tag{3}$$

where L^{-1} denotes the inverse Laplace transform. Using tables of Laplace transforms, we obtain

$$f_T(t) = \begin{cases} \dfrac{\alpha^r}{\Gamma(r)} t^{r-1} e^{-\alpha t}, & t \geq 0, \alpha > 0, r > 0, \\ 0, & t < 0. \end{cases} \tag{4}$$

The distribution defined by (4) is called the gamma distribution. The manner in which we have obtained it shows it to be a natural generalization of the exponential distribution. When $r = 1$, it reduces to the exponential distribution. The gamma distribution is a distribution in its own right and we have only obtained it here as a generalization of the exponential distribution to establish it in a logical way as a physical model. For

(4) to be a pdf it is not necessary that r be an integer, and hence we drop that requirement and simply insist that $r > 0$.

Although we have assumed that each of the components contributing to the instantaneous failure rate of T was constant, the hazard function for the gamma distribution is not constant. By definition

$$H(t) = \frac{\alpha^r t^{r-1} e^{-\alpha t}}{\Gamma(r) - \alpha^r \int_0^t x^{r-1} e^{-\alpha x}\, dx}. \tag{5}$$

Equation (5) cannot be written in a closed form unless r is a positive integer. However,

$$\begin{aligned}\Gamma(r) F_T(t) &= \alpha^r \int_0^t x^{r-1} e^{-\alpha x}\, dx \\ &= \int_0^t (\alpha x)^{r-1} e^{-\alpha x}\, \alpha\, dx \\ &= \int_0^{\alpha t} y^{r-1} e^{-y}\, dy,\end{aligned}$$

and the latter quantity, called the incomplete gamma function, has been extensively tabulated. See Harter (Ref. 11). Using these tables one can calculate values of the cumulative distribution $F_T(t)$ for a wide range of t. $H(t)$ can then be obtained from (5). Figure 4-11 shows $H(t)$ for $r = .5$, 1.0, and 1.5 and $\alpha = 1$.

The quantities α and r in the gamma distribution are known as the scale and shape parameters, respectively. The reason for these names is clear since we can see from Figures 4.12 and 4.13 that α only effects the scale of the curve, whereas r varies its general shape.

The parameters α and r are related to the mean and variance of the gamma distribution. By (2) we have

$$\mu_X = \frac{d}{d\theta}\left(\frac{\alpha}{\alpha - \theta}\right)^r \Bigg|_{\theta=0} = \frac{r}{\alpha} \tag{6}$$

and

$$\sigma_X^2 = \frac{d^2}{d\theta^2}\left(\frac{\alpha}{\alpha - \theta}\right)^r \bigg|_{\theta=0} - \frac{r^2}{\sigma^2} = \frac{r}{\alpha^2}. \tag{7}$$

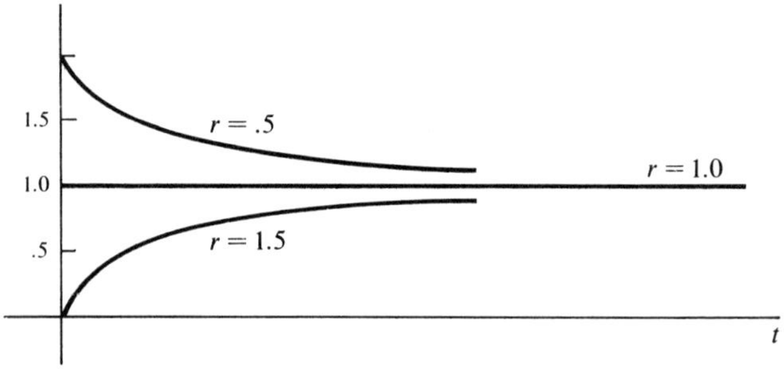

Figure 4.11 Hazard Functions for the Gamma Distribution with $\alpha = 1$

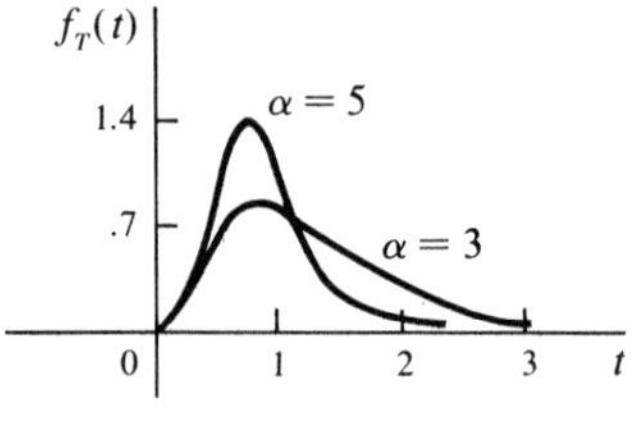

Figure 4.12 Gamma pdf's with $r = 3$

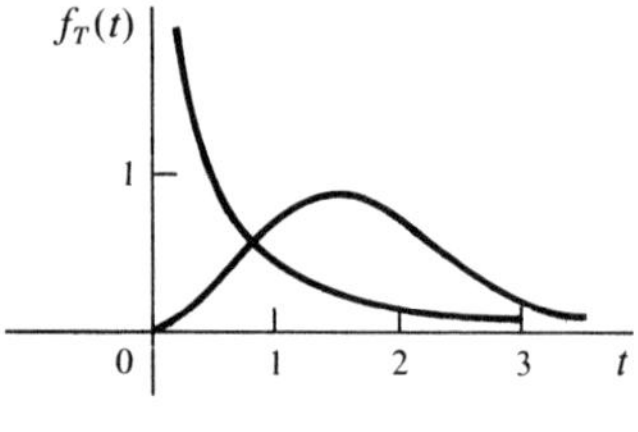

Figure 4.13 Gamma pdf's with $\alpha = 1$

4.16 THE WEIBULL DISTRIBUTION

One of the major criticisms leveled at the exponential distribution as a time-to-failure distribution is the fact that its hazard function is constant. A distribution which escapes this criticism and which encompasses the exponential distribution arises from the hazard function,

$$H(t) = \frac{\alpha}{\lambda}\left(\frac{t}{\lambda}\right)^{\alpha-1}. \tag{1}$$

By solving (3), Section 4.13, with $H(t)$ given by (1) we obtain

$$f_T(t) = \begin{cases} \dfrac{\alpha}{\lambda}\left(\dfrac{t}{\lambda}\right)^{\alpha-1} \exp\left[-\left(\dfrac{t}{\lambda}\right)^{\alpha}\right], & t \geq 0, \alpha > 0, \lambda > 0, \\ 0, & t < 0. \end{cases} \tag{2}$$

The distribution defined by (2) is called the *Weibull distribution.* It has been posed as a probability model for time to failure on the basis of (2) and empirical studies. One of its most frequent uses has been in studies on the life length of electron tubes. Some plots of the distribution and its hazard function are given in Figures 4.14 and 4.15.

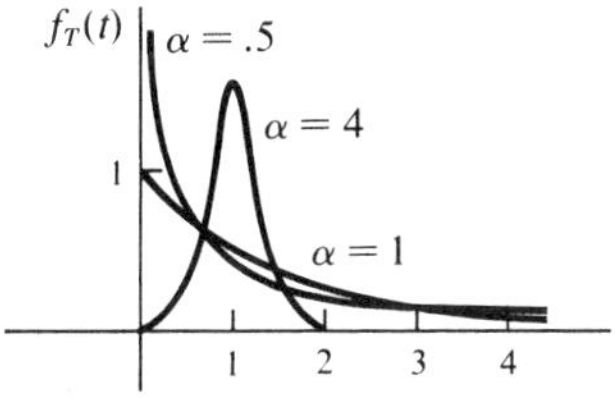

Figure 4.14 Weibull pdf's with $\lambda = 1$

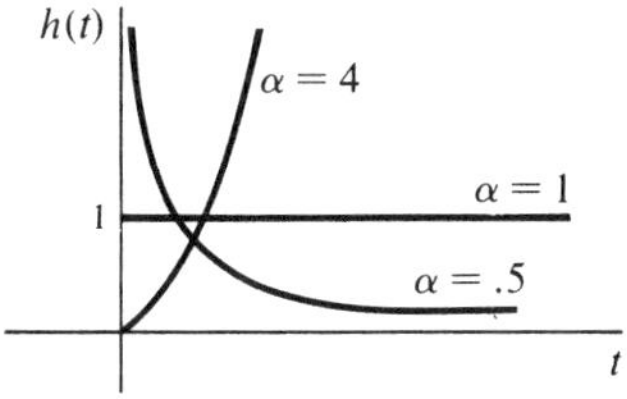

Figure 4.15 Hazard Functions for the Weibull Distribution with $\lambda = 1$

4.17 THE LOG NORMAL DISTRIBUTION

If X is a random variable whose logarithm has a normal distribution, then we say that X has a log normal distribution. If $E[\ln X] = \alpha$ and $\text{Var}[\ln X] = \beta^2$, that is, $\ln X \sim N(\alpha, \beta^2)$, then by the methods of Chapter II we obtain for the pdf f_X of a log normal random variable

$$f_X(x) = \begin{cases} \dfrac{1}{\beta X\sqrt{2\pi}} \exp\left[-\dfrac{1}{2\beta^2}(\ln X - \alpha) \right], & X > 0, \\ 0, & X \leq 0. \end{cases} \tag{1}$$

Owing to its wide variety of shapes, the distribution has found extensive application in a number of areas.

It has been used considerably in reliability as a time-to-failure model and has also been used as a time-to-repair model. For a complete discussion of the distribution, its properties and applications, see Aitchison and Brown (Ref. 1). Plots of thc distribution and its hazard function are shown in Figures 4.16 and 4.17.

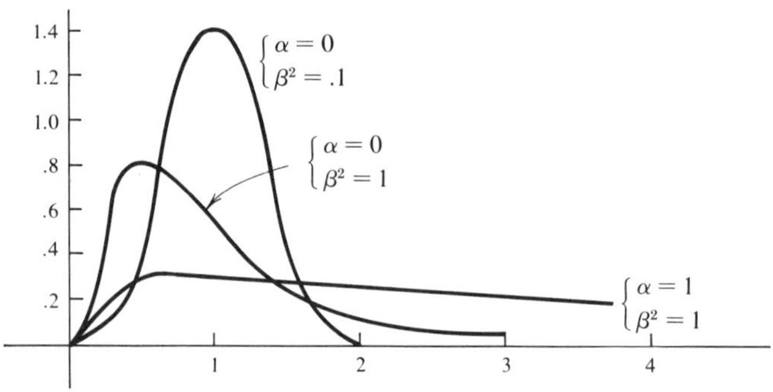

Figure 4.16 Log Normal Distributions

4.18 THE UNIFORM DISTRIBUTION

If a discrete random variable X takes on each of the n possible values $x_1, \ldots, x_n$ with equal probability $1/n$, that is, $P[X = x_i] = 1/n$ for each i, then we say it has a discrete uniform distribution. The continuous analogue of this is as follows. If X is a continuous random variable and

$$f_X(x) = \begin{cases} \dfrac{1}{\beta - \alpha}, & \text{if } \alpha < x < \beta, \\ 0, & \text{elsewhere,} \end{cases} \tag{1}$$

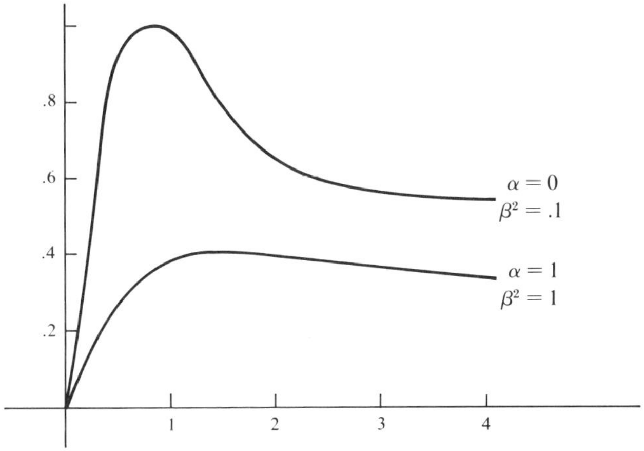

Figure 4.17 Hazard Function for Log Normal Distribution

then we say that X is uniformly distributed over the interval (α, β) or $X \sim U(\alpha, \beta)$. The distribution defined by (1) is also often referred to as the rectangular distribution. The rectangular or uniform distribution is an appropriate model when the probability that X has a value within a given finite interval is a function only of the interval length. For example, suppose a signal arrives at random in a 3-minute interval and the time of arrival x is equally likely for every x such that $0 < x \leq 3$. Then the probability that the signal arrives during any time interval is a function only of its length. Thus the $U(0, 3)$ distribution is an appropriate probability model, and the probability that the signal will arrive in the first minute is given by

$$P[X \leq 1] = \tfrac{1}{3} \int_0^1 dx = \tfrac{1}{3}.$$

The moments of all orders are quite easily obtained from equation (1) and

$$\mu = \frac{1}{\beta - \alpha} \int_\alpha^\beta x\,dx = \frac{\alpha + \beta}{2}, \tag{2}$$

$$\sigma^2 = \frac{1}{\beta - \alpha} \int_\alpha^\beta x^2\,dx - \left(\frac{\alpha + \beta}{2}\right)^2$$

$$= \frac{(\beta - \alpha)^2}{12}. \tag{3}$$

One of the more important properties of the uniform distribution is given by the following theorem.

Theorem 1 Let X be a continuous random variable and let $Y = F_X(x)$, where F_X is the cumulative distribution of X. Then $Y \sim U(0, 1)$; that is,

$$F_Y(y) = \begin{cases} 1, & 0 < y < 1, \\ 0, & \text{elsewhere.} \end{cases}$$

The above theorem states that any continuous random variable can be transformed to a random variable with a $U(0, 1)$ distribution. We will now exemplify this theorem.

Example 1. Let $X \sim e^{-x}$, $x > 0$. Find a transformation on x which transforms X to a $U(0, 1)$ random variable.

Solution: By the above theorem if we let

$$y = \int_0^x e^{-t}\, dt = 1 - e^{-x}, \tag{4}$$

then the random variable $Y \sim U(0, 1)$. To verify this in another way we apply Theorem 1, Section 2.8.

Since from (4), $x = \ln(1 - y)^{-1}$, we have

$$\begin{aligned} f_Y(y) = f_X(x)\left|\frac{dx}{dy}\right| &= e^{\ln(1-y)} \cdot \frac{1}{1-y} \\ &= 1, \qquad 0 < y < 1, \end{aligned}$$

as asserted by Theorem 1.

Example 2. Let $Y \sim U(0,1)$. Find a transformation which will transform Y to a random variable X such that

$$f_X(x) = \begin{cases} e^{-x}, & x > 0, \\ 0, & x \leq 0. \end{cases}$$

Solution: To solve this we simply take the inverse transformation of the one used in Example 1. Thus we let $x = \ln(1 - y)^{-1}$. Then

$$y = 1 - e^{-x}$$

and

$$f_X(x) = \begin{cases} f_Y(y)\left|\dfrac{dy}{dx}\right| = e^{-x}, & x > 0, \\ 0, & x \le 0. \end{cases}$$

Example 2 illustrates an important point. That is, by inverting the transformation which transforms a given random variable to one with a uniform distribution, we obtain a transformation which produces the original variate from a uniform variate. This result is of some consequence in random-number generation, since most computers have a $U(0, 1)$ random-number generator available. Theoretically, one can generate from these numbers any random numbers he desires simply by solving for x in the equation

$$y = F_X(x). \tag{5}$$

This is precisely what was done in Example 2. That is, given any y from a $U(0, 1)$ distribution we showed that

$$x = \ln(1 - y)^{-1}$$

was a random number from an exponential distribution. Of course, in many cases (5) may be difficult to solve analytically and numerical methods may be necessary. In still other cases this may not be the best way to generate the desired random numbers. It is, however, a general method and is of some practical as well as theoretical value.

4.19 THE MULTIVARIATE NORMAL DISTRIBUTION

Let $X_1, \ldots, X_n$ be random variables with the joint pdf $f_{X_1 \cdots X_n}$ defined by

$$f_{X_1 \cdots X_n}(x_1, \ldots, x_n)$$

Table 4.2

Density Function	Functional Description	Mean	Variance	Moment-Generating Function
Normal	$f_X(x) = \frac{1}{\sigma\sqrt{2\pi}} \exp\left[-\frac{1}{2}\left(\frac{X - \mu}{\sigma}\right)^2\right], \quad -\infty < x < \infty.$	μ	σ^2	$\exp\left(\mu t + \frac{1}{2}\sigma^2 t^2\right)$
Chi square	$f_X(x) = \begin{cases} \frac{x^{n/2}e^{-(1/2)x}}{\Gamma(n/2)2^{n/2}}, & x \geq 0, \\ 0, & x < 0 \end{cases}$	n	$2n$	$(1 - 2t)^{-n/2}$
Cauchy	$f_X(x) = \frac{\alpha}{\pi} \frac{1}{\alpha^2 + (x - \mu)^2}, \quad -\infty < x < \infty.$	Exists only as the principal value μ	Does not exist	
Exponential	$f_X(x) = \begin{cases} \alpha e^{-\alpha x}, & x \geq 0, \\ 0, & x < 0 \end{cases}$	$\frac{1}{\alpha}$	$\frac{1}{\sigma^2}$	$\frac{\alpha}{\alpha - t}$
Gamma	$f_X(x) = \begin{cases} \frac{\alpha}{r} x^{r-1} e^{-\alpha x}, & x \geq 0, \\ 0, & x < 0 \end{cases}$	$\frac{r}{\alpha}$	$\frac{r}{\alpha^2}$	$\left(\frac{\alpha}{\alpha - t}\right)^r$
Uniform	$f_X(x) = \begin{cases} \frac{1}{\beta - \alpha}, & \alpha < x < \beta, \\ 0, & \text{elsewhere} \end{cases}$	$\frac{\alpha + \beta}{2}$	$\frac{(\beta - \alpha)^2}{12}$	$\frac{1}{t}\left(\frac{e^{\beta t} - e^{\alpha t}}{\beta - \alpha}\right), t \neq 0.$

$$= \frac{|R|^{1/2}}{(2\pi)^{n/2}} \exp\left[-\tfrac{1}{2} \sum_{j=1}^{n} \sum_{i=1}^{n} (x_i - \mu_i)(x_j - \mu_j) r_{ij}\right],$$

$$-\infty < x_j < \infty, \quad j = 1, 2, \ldots, n, \qquad (1)$$

where $|R|$ is the determinant whose elements are the r_{ij}, and the r_{ij} are parameters such that

$$\sum_{j=1}^{n} \sum_{i=1}^{n} (x_i - \mu_i)(x_j - \mu_j) r_{ij} > 0.$$

Then the random variables $X_1, \ldots, X_n$ are said to have a multivariate normal distribution. An investigation of this important joint density function is beyond the scope of our discussion. However, by making use of the theory of matrices the distribution is easy to handle. We mention it here for reference purposes only. It is the most extensively studied of all multivariate distributions. See Anderson (Ref. 2).

V
Statistical Estimation

5.1 INTRODUCTION

One of the most important applications of probability theory is in providing a precise basis for the theory of statistical estimation. The central problem in statistical estimation can be stated as follows. By observing a set of values of the random variable X, whose probability density function is defined by

$$f_X(x; \theta),$$

where θ is an unknown parameter, what is the best way to estimate θ? For example, suppose one wishes to determine the probability that a head will appear on a toss of a particular coin. Let X be a random variable such that

$$x = \begin{cases} 1, & \text{if a head appears,} \\ 0, & \text{if a tail appears.} \end{cases}$$

Let θ denote the probability of obtaining a head on a single toss. Then

$$f_X(x; \theta) = \theta^x (1 - \theta)^{1-x}, \qquad x = 0, 1$$

is the probability density function of the random variable X.

Note that θ is not known. Yet by tossing the coin several times, say n times, one feels that a "good estimate" of θ would be

$$\overline{x} = \sum_{i=1}^{n} \frac{x_i}{n}, \tag{1}$$

that is, the proportion of times that heads appear out of n tries.

Clearly, (1) could have different values for each distinct set of n tosses and is, therefore, best thought of as a value of a random variable. Such random variables are referred to as *estimators*. That is, a random variable whose values are used as estimates for unknown parameters are called estimators. The particular estimator we are considering here is the sample mean $\overline{X}$ discussed in Chapter III. An estimator for θ based on a sample $X_1, \ldots, X_n$ is, therefore, a random variable $T(X_1, \ldots, X_n)$ whose values are completely determined by $x_1, \ldots, x_n$.

Since virtually any random variable could serve as an estimator, it behooves us to characterize those estimators which are logically more desirable than others. This topic, that is, the problem of determining criteria which an estimator should satisfy, would lead us much deeper into the field of mathematical statistics than we wish to go. We will mention only a few such criteria. The reader who wishes a more complete discussion should see Chapter VIII of Graybill (Ref. 10).

5.2 UNBIASED ESTIMATORS

If $\hat{\theta}$ is an estimator for θ such that

$$E[\hat{\theta}] = \theta, \tag{1}$$

then we call $\hat{\theta}$ an unbiased estimator for θ.

By Example 3, Section 3.3, $E[\overline{X}] = \mu_X$ and hence $\overline{X}$ is always an unbiased estimator for μ_X. Speaking rather loosely, this means that if an infinite collection of values of $\overline{X}$ could be obtained they would average to μ_X.

Although the sample mean is an unbiased estimator for μ_X, the sample variance S^2 is not an unbiased estimator for σ_X^2. This is seen as follows. Let $X_1, \ldots, X_n$ be a random sample of size n. Then

$$E[S^2] = E\left[\frac{1}{n}\sum_{i=1}^{n}(X_i - \overline{X})^2\right]$$

$$= E\left\{\frac{1}{n}\sum_{i=1}^{n}[(X_i - \mu_X) - (\overline{X} - \mu_X)]^2\right\}$$

$$= \frac{1}{n} \sum_{i=1}^{n} \{E[(X_i - \mu_X)^2]$$
$$-2E[(X_i - \mu_X)(\overline{X} - \mu_X)] + E[(\overline{X} - \mu_X)^2]\}$$
$$= \sigma_X^2 - \frac{2}{n} E\left\{\sum_{i=1}^{n} (\overline{X} - \mu_X)(X_i - \mu_X)\right\} + \sigma_{\overline{X}}^2$$
$$= \sigma_X^2 - \frac{2}{n} E\left\{(\overline{X} - \mu_X) \sum_{i=1}^{n} (X_i - \mu_X)\right\} + \frac{\sigma_X^2}{n}$$
$$= \sigma_X^2 - 2E[(\overline{X} - \mu)^2] + \frac{\sigma_X^2}{n}$$
$$= \sigma_X^2 - 2\sigma_{\overline{X}}^2 + \frac{\sigma_X^2}{n}$$
$$= \sigma_X^2 - \frac{2\sigma_X^2}{n} + \frac{\sigma_X^2}{n}$$
$$= \frac{n-1}{n} \sigma_X^2. \tag{2}$$

From (2) it is clear that S^2 is a biased estimator for σ_X^2. As can also be seen from (2), this bias is negligible for large n, but is quite serious for small values of n. Note that

$$E\left[\frac{n}{n-1} S^2\right] = \frac{n}{n-1} E[S^2]$$
$$= \frac{n}{n-1} \frac{n-1}{n} \sigma_X^2$$
$$= \sigma_X^2,$$

so that

$$\frac{n}{n-1} S^2 = \frac{1}{n-1} \sum_{i=1}^{n} (X_i - \overline{X})^2 \tag{3}$$

is an unbiased estimator for σ_X^2. For this reason some authors define the sample variance by (3) rather than the form we have chosen. In regard to this, we should make the comment that the property of being unbiased is only one of many properties which one might deem important for an estimator to possess. Often it is not possible to find a single estimator which exhibits all these properties, and hence in correcting for one property we may lose another property. This is true in the case of S^2. By correcting for bias and hence using (3) as our estimator for σ_X^2, we find in Section 5.4 that we lose the maximum likelihood property in the important normal case. Thus the engineer should consider such corrections closely in light of the sample size and purpose of a proposed estimator.

5.3 MINIMUM VARIANCE ESTIMATORS

A popular and useful criterion for judging the quality of unbiased estimators is the so-called *minimum variance criterion*. Let θ^* be an element of the set of all unbiased estimators for θ; then $\theta^* = \hat{\theta}$ is said to be a minimum variance unbiased estimator for θ if

$$E(\hat{\theta} - \theta)^2 \leq E(\theta^* - \theta)^2$$

for every θ^* in the set of all unbiased estimators for θ. If $\hat{\theta}$ is that estimator such that strict inequality holds, then $\hat{\theta}$ is the unique minimum variance unbiased estimator for θ. If $\hat{\theta}$ is an unbiased minimum variance estimator we will refer to it as a best estimator. It is important to note that these estimators may not exist in every case, and when they do exist they are in general difficult to find.

Let $X_1, \ldots, X_n$ be a random sample with pdf defined by $f_X(x;\mu)$ and let $L(x;\mu) = f_X(x_1;\mu)\cdots f_X(x_n;\mu)$. Since $L(x;\mu)$ is a joint pdf we know that

$$1 = \int_{-\infty}^{\infty} \cdots \int_{-\infty}^{\infty} L(x;\mu)\, dx_1 \cdots dx_n. \tag{1}$$

Now assume that the following regularity conditions hold:

$$\text{(a)}\ \frac{\partial L}{\partial \mu}\left[\int_{-\infty}^{\infty} \cdots \int_{-\infty}^{\infty} L(x;\mu)\, dx_i \cdots dx_n\right]$$

$$= \int_{-\infty}^{\infty} \cdots \int_{-\infty}^{\infty} \frac{\partial L}{\partial \mu}\, dx_1 \cdots dx_n$$

and

$$\text{(b)}\quad \frac{\partial L}{\partial \mu}\left[\int_{-\infty}^{\infty}\cdots\int_{-\infty}^{\infty} T(x_1,\ldots,x_n)L\,dx_1\cdots dx_n\right] = \int_{-\infty}^{\infty}\cdots\int_{-\infty}^{\infty} T(x_1,\ldots,x_n)\,\frac{\partial L}{\partial \mu}\,dx_1\cdots dx_n,$$

where

$$T = T(X_1,\ldots,X_n)$$

is an unbiased estimator for μ.

On differentiating both sides of (1) with respect to μ and by condition (a),

$$0 = \int_{-\infty}^{\infty}\cdots\int_{-\infty}^{\infty} \frac{\partial L(x;\mu)}{\partial \mu}\,dx_1\cdots dx_n$$

or

$$0 = \int_{-\infty}^{\infty}\cdots\int_{-\infty}^{\infty} \frac{1}{L(x;\mu)}\,\frac{\partial L(x;\mu)}{\partial \mu}\,L(x;\mu)\,dx_1\cdots dx_n = E\left[\frac{\partial \ln L(X;\mu)}{\partial \mu}\right]. \tag{2}$$

Now we define the random variable

$$S = \frac{\partial \ln L(X;\mu)}{\partial \mu}.$$

Then

$$E(S) = 0$$

and

$$\sigma_S^2 = E\left[\frac{\partial \ln L(X;\mu)}{\partial \mu}\right]^2.$$

Since T is an unbiased estimator for μ, then

$$\mu = \int_{-\infty}^{\infty}\cdots\int_{-\infty}^{\infty} TL(x;\mu)\,dx_1\cdots dx_n. \tag{3}$$

On differentiating with respect to μ both sides of (3) we obtain

$$1 = \int_{-\infty}^{\infty}\cdots\int_{-\infty}^{\infty} T\,\frac{\partial L(x;\mu)}{\partial \mu}\,dx_1\cdots dx_n$$

$$= \int_{-\infty}^{\infty} \cdots \int_{-\infty}^{\infty} T\left(\frac{1}{L(x;\mu)}\right)\frac{\partial L(x;\mu)}{\partial \mu} L(x;\mu)\, dx_1 \cdots dx_n$$

$$= \int_{-\infty}^{\infty} \cdots \int_{-\infty}^{\infty} T\, \frac{\partial \ln L(x;\mu)}{\partial \mu} L(x;\mu)\, dx_1 \cdots dx_n$$

$$= \int_{-\infty}^{\infty} \cdots \int_{-\infty}^{\infty} (TS)L(x;\mu)\, dx_1 \cdots dx_n$$

or

$$1 = E[TS] = \sigma_{TS},$$

the covariance of T and S.

Let the variance of T be denoted as σ_T^2. Now for any real number k,

$$\begin{aligned} 0 &\le E[(T - \mu) + kS]^2 \\ &\le E[(T - \mu)^2 + 2k(T - \mu)S + k^2S^2] \\ &\le \sigma_T^2 + 2k\sigma_{TS} + \sigma_S^2k^2. \end{aligned} \tag{4}$$

But the right side of (4) is a quadratic in k which is nonnegative for all values of k; that is, it has at most one distinct real root. Thus

$$\sigma_{TS}^2 - \sigma_T^2\sigma_S^2 \le 0,$$

or

$$\sigma_T^2\sigma_S^2 \ge \sigma_{TS}^2$$

and

$$\sigma_T^2 \ge \frac{\sigma_{TS}^2}{\sigma_S^2},$$

which is a lower bound for the variance of T, any statistic that is an unbiased estimator for μ.

But $\sigma_{TS}^2 = 1$, and the bound is simply

$$\sigma_T^2 \ge \frac{1}{\sigma_S^2} \tag{5}$$

Example 1. To exemplify the above, let $X \sim N(\mu, \sigma^2)$. Then

$$L(x; \mu) = \frac{1}{(2\pi)^{n/2}} \exp\left[-\tfrac{1}{2}\,\sigma^2 \sum_{i=1}^{n} (x_i - \mu)^2\right],$$

$$\ln L(x; \mu) = -\frac{n}{2} \ln 2\pi - \frac{1}{2\sigma^2} \sum_{i=1}^{n} (x_i - \mu)^2,$$

$$\begin{aligned} S &= \partial \ln L(x; \mu)/\partial \mu \\ &= \frac{1}{\sigma^2} \sum_{i=1}^{n} (x_i - \mu) \\ &= \frac{n}{\sigma^2} (\bar{X} - \mu). \end{aligned}$$

Since variance of $(\bar{X} - \mu)$ is σ^2/n, it follows that

$$\sigma_S^2 = \frac{n^2}{\sigma^4} \frac{\sigma^2}{n} = \frac{n}{\sigma^2}.$$

Thus from (3) we see that

$$\sigma_T^2 \geq \frac{\sigma^2}{n} \tag{6}$$

for any unbiased estimator T. However,

$$\sigma_{\bar{X}}^2 = \frac{\sigma^2}{n},$$

which, by (5), is the lower bound for any estimator of μ, and thus $\bar{X}$ is a best estimator for μ, that is, minimum variance unbiased estimator.

In the above example, it was necessary to assume that $X \sim N(\mu, \sigma^2)$. This assumption can be eliminated if we restrict ourselves to unbiased linear estimators. If $X_1, \ldots, X_n$ is a random sample, we say that an estimator is linear if it is a weighted sum of the random variables $X_1, \ldots, X_n$. Let the weights be n constants $a_1, a_2, \ldots, a_n$, which we will select judiciously so that the resulting estimator is unbiased and minimum variance. All linear estimators, therefore, are defined to be of the form

$$\mu^* = \sum_{i=1}^{n} a_i X_i,$$

where $X_1, \ldots, X_n$ is a random sample. The condition that μ^* is unbiased implies that

$$\begin{aligned} \mu &= E(\mu^*) \\ &= E\left(\sum_{i=1}^{n} a_i X_i\right) \\ &= \sum_{i=1}^{n} a_i E(X_i) \\ &= \sum_{i=1}^{n} a_i \mu, \end{aligned}$$

or

$$\sum_{i=1}^{n} a_i = 1. \tag{7}$$

Let us represent each a_i as

$$a_i = \frac{c_i}{\sum_{j=1}^{n} c_j}, \qquad j = 1,2,\ldots,n,$$

where the c_i's are constants. This assures us that (7) is true and allows us now to search for each c_i to obtain the minimum variance estimator $\hat{\mu}$ for μ. Let

$$\mu^* = \frac{\sum_{i=1}^{n} c_i X_i}{\sum_{j=1}^{n} c_j}.$$

Since the sample is a random sample, the variance, V, of μ^* is simply

$$V = \sum_{i=1}^{n} \frac{c_i^2}{(\Sigma c_j)^2}\, \sigma^2,$$

where the variance of each X_i is σ^2. A necessary condition for V to be minimal is

$$\frac{\partial V}{\partial c_k} = 0, \qquad k = 1,2,\ldots,n.$$

This requirement, which turns out to also be sufficient, gives n equations in n unknowns to determine the c_i's:

$$\begin{aligned}\frac{\partial V}{\partial c_k} &= \sigma^2 \frac{d}{dc_k} \frac{\sum_{i=1}^{n} c_i^2}{(\Sigma c_j)^2} \\ &= \frac{2(\Sigma c_j)\Sigma c_i^2 - (\Sigma c_j)^2 \cdot 2c_k}{(\Sigma c_j)^4} \\ &= 0\end{aligned}$$

implies that

$$c_k = \frac{\Sigma c_i^2}{\Sigma c_j}, \qquad \text{for all } k.$$

Consequently,

$$c_1 = c_2 = \cdots = c_n = c$$

for some constant c, which in turn implies that

$$\begin{aligned}a_i &= \frac{c}{nc} \\ &= \frac{1}{n}.\end{aligned}$$

Thus we conclude that

$$\hat{\mu} = \sum_{i=1}^{n} \frac{1}{n} X_i,$$

or simply

$$\hat{\mu} = \overline{X}$$

is the minimum variance *linear* unbiased estimator for $E(X) = \mu$ regardless of the probability density function of X.

5.4 MAXIMUM LIKELIHOOD ESTIMATION

Consider the following hypothetical problem. Suppose it is known that an urn contains either eight red balls and two white balls or two red balls and eight white balls. Suppose further that three balls are drawn successively with replacement and they are all red. What is the most logical estimate, based on this sample, for the probability p of drawing a red ball in a single trial? It seems quite obvious here that the urn most likely contains eight red balls and two white balls, and hence the most logical estimate for p, $\hat{p}$, is

$$\hat{p} = \tfrac{8}{10}.$$

When we say "most logical estimate" we are intuitively reasoning that drawing three red balls in a row seems more support for the conclusion that $p = \frac{8}{10}$ than $p = \frac{2}{10}$. This is true in the sense that

$$P[3 \text{ red} | p = \tfrac{2}{10}] = (\tfrac{2}{10})^3 = \tfrac{1}{125},$$

while

$$P[3 \text{ red} | p = \tfrac{8}{10}] = (\tfrac{8}{10})^3 = \tfrac{64}{125}.$$

Thus we have selected from all possible choices of p the value for p which makes our observed outcome the most probable.

Note that if X is a random variable which assigns the number 1 to red and 0 to white, then

$$P[X = x_i] = \begin{cases} p^{x_i}(1 - p)^{1 - x_i}, & x_i = 0, 1, \\ 0, & \text{elsewhere}, \end{cases} \tag{1}$$

where $p = \frac{2}{10}$ or $\frac{8}{10}$.

Consequently, in the above we have estimated the parameter p in the pdf defined by (1). The procedure which we have used is called maximum likelihood estimation and the estimate which results is called a maximum likelihood estimate. We will now formalize these ideas.

Definition 1 Let $X_1, X_2, \ldots, X_n$ be a random sample with pdf defined by $f_X(x; \theta)$, where θ belongs to some set S, i.e., $\theta \epsilon S$. Then we define the likelihood function L by

$$L(x_1, \ldots, x_n; \theta) = f_X(x_1; \theta) \cdots f_X(x_n; \theta), \tag{2}$$

and the value of θ, $\hat{\theta}$, which maximizes L, we call the *maximum likelihood estimate* of θ.

Example 1. Let

$$f_X(x; \theta) = \begin{cases} \theta^x(1 - \theta)^{1-x}, & 0 < \theta < 1, x = 0, 1, \\ 0, & \text{elsewhere.} \end{cases} \tag{3}$$

Suppose a random sample from the distribution defined by (3) is observed. What is the maximum likelihood estimate of θ?

Solution: In this case,

$$L(x_1, x_2, \ldots, x_n; \theta) = \theta^{x_i}(1 - \theta)^{1-x_i} \cdots \theta^{x_n}(1 - \theta)^{1-x_n}$$

$$= \theta^{\sum_{i=1}^{n} x_i} \Big/ (1 - \theta)^{n - \sum_{i=1}^{n} x_i}$$

A necessary condition for $L(x_1, \ldots, x_n)$ to be minimal is that the derivative $dL/d\theta = 0$, or

$$\Sigma(x_i)\theta^{\Sigma x_i - 1}(1 - \theta)^{n - \Sigma x_i} - (n - \Sigma x_i)\theta^{\Sigma x_i}(1 - \theta)^{n - \Sigma x_i - 1} = 0$$

or

$$\theta^{\Sigma x_i - 1}(1 - \theta)^{n - \Sigma x_i - 1}[\Sigma x_i(1 - \theta) - (n - \Sigma x_i)\theta] = 0.$$

Since $\theta \neq 0$ or 1,

$$\Sigma x_i(1 - \theta) - (n - \Sigma x_i)\theta = 0$$

or

$$\Sigma x_i - \theta \Sigma x_i - n\theta + \Sigma x_i \theta = 0.$$

And finally, solving for θ we find that the maximum likelihood estimate for θ is

$$\hat{\theta} = \frac{\Sigma x_i}{n}.$$

One can interpret this result in the following manner: Given the observations $x_1, \ldots, x_n$, the density function with the parameter $\theta = \hat{\theta}$ is the one "most likely" to have produced the sample.

Example 2. Let us consider a second probability density function, this time with two parameters, μ and σ^2. A useful continuous probability density function of this type is the normal, which as we know can be written

$$f_X(x; \mu, \sigma^2) = (2\pi)^{-1/2}(\sigma^2)^{-1/2} \exp\left[-\frac{1}{2}\frac{(x-\mu)^2}{\sigma^2}\right].$$

Let $X_1, X_2, \ldots, X_n$ denote a random sample. The likelihood function is then

$$L(x_1, \ldots, x_n; \theta) = (2\pi)^{-n/2}(\sigma^2)^{-n/2} \exp\left[-\frac{1}{2}\sigma^2 \sum_{i=1}^{n} (x_i - \mu)^2\right].$$

The maximum likelihood estimates of μ and σ^2 result from the equations

$$\frac{\partial L}{\partial \mu} = 0$$

and

$$\frac{\partial L}{\partial \sigma^2} = 0.$$

The first equation,

$$\frac{\partial L}{\partial \mu} = \frac{1}{\sigma^2}\sum_{i=1}^{n} (x_i - \mu) = 0$$

yields the maximum likelihood estimate $\hat{\mu}$ for μ, where

$$\hat{\mu} = \sum_{i=1}^{n} \frac{x_i}{n},$$

the arithmetic mean of the sample values. The second equation,

$$\frac{\partial L}{\partial \sigma^2} = -\frac{n}{2}\sigma^{-4} + \frac{1}{2\sigma^2}\sum_{i=1}^{n} (x_i - \mu)^2,$$

along with the fact that

$$\hat{\mu} = \bar{x}$$

yields

$$\hat{\sigma}^2 = \frac{1}{n}\sum_{i=1}^{n} (x_i - \bar{x})^2.$$

Clearly, $\hat{\sigma}^2$ is a value of S^2, the sample variance. Thus the sample variance is a maximum likelihood estimator for σ^2 in the normal case. As remarked earlier, it is not, however, an unbiased estimator.

Example 3. Let

$$x_1 = 5, \quad x_2 = 3, \quad x_3 = 1, \quad x_4 = 4, \quad x_5 = 2$$

represent a random sample from a normal population with mean μ and variance σ^2. Find a point estimate for μ and for σ^2.

Solution: The maximum likelihood estimate for μ is

$$\begin{aligned} \bar{x} &= \sum_{i=1}^{n} \frac{x_i}{n} \\ &= \frac{5 + 3 + 1 + 4 + 2}{5} \\ &= 3. \end{aligned}$$

The maximum likelihood estimator for σ^2 is

$$\begin{aligned} \hat{\sigma}^2 &= \frac{\Sigma(x_i - x)^2}{n} \\ &= \frac{4 + 0 + 4 + 1 + 1}{5} \\ &= 2. \end{aligned}$$

Example 4. Let $X_1, \ldots, X_n$ be a random sample from the distribution defined by

$$f_X(x;\theta) = \begin{cases} \dfrac{1}{\theta}, & 0 \le x \le 0, \\ 0, & \text{elsewhere.} \end{cases}$$

Find the maximum likelihood estimate for θ.

Solution: The likelihood function is given by

$$L(x_1, \ldots, x_n; \theta) = f_X(x_1; \theta) \cdots f_X(x_n; \theta),$$
$$= \begin{cases} \theta^{-n}, & 0 < x_i < \theta, i = 1, 2, \ldots, n, \\ 0, & \text{elsewhere.} \end{cases}$$

However, if $\hat{\theta} < \max\{x_1, \ldots, x_n\}$, then $x_k > \hat{\theta}$ for some k and, therefore, $f_X(x_k; \hat{\theta}) = 0$, so that $L(x_1, \ldots, x_n; \hat{\theta}) = 0$. Thus $\hat{\theta} \geq \max\{x_1, \ldots, x_n\}$. But in this event

$$L(x_1, \ldots, x_n; \hat{\theta}) = \frac{1}{\hat{\theta}^n} \tag{4}$$

and clearly (4) is a maximum if $\hat{\theta}$ is a minimum; that is,

$$\hat{\theta} = \max\{x_1, \ldots, x_n\}. \tag{5}$$

Example 5. Let $x_1 = 2, x_2 = 6, x_3 = 3, x_4 = 1$, and $x_5 = 4$ be observed values of a random sample from the distribution of Example 4, that is, a $U(0, \theta)$ distribution. The maximum likelihood estimate for θ is then

$$\hat{\theta} = \max\{x_1, \ldots, x_5\}$$
$$= 6.$$

5.5 INTERVAL ESTIMATORS

Until now we have only discussed point estimators, that is, one dimensional estimators or estimators which produce a single number as an estimate. This is, of course, not necessary or even always desirable. It may be more reasonable to estimate θ by an interval than a point, in which case we may estimate θ to be, say, somewhere in the interval (a, b). In this event our estimation problem is one of determining a and b in a logical manner from the observed data and our estimator is then referred to as a two-dimensional or interval estimator.

Example 1. Let $X_1, \ldots, X_n$ be a random sample from a $N(\mu, \sigma^2)$ distribution. Then

$$\frac{\overline{X} - \mu}{\sigma/\sqrt{n}} \sim N(0, 1).$$

Suppose σ^2 is known; then

$$P\left[-k_{\alpha/2} \leq \frac{\bar{X} - \mu}{\sigma/\sqrt{n}} \leq k_{\alpha/2}\right] = 1 - \alpha, \tag{1}$$

where $k_{\alpha/2}$ is selected from Table I such that

$$\int_{-k_{\alpha/2}}^{k_{\alpha/2}} \frac{1}{(2\pi)^{1/2}} e^{-(1/2)t^2}\, dt = 1 - \alpha.$$

Thus $k_{\alpha/2}$ is known. By elementary manipulation of (1) we can rewrite it in the form

$$P\left[\bar{X} - \frac{\sigma}{\sqrt{n}} k_{\alpha/2} \leq \mu \leq \bar{X} + \frac{\sigma}{\sqrt{n}} k_{\alpha/2}\right] = 1 - \alpha. \tag{2}$$

But equation (2) defines a random interval or two-dimensional estimator for μ: the random interval $(\bar{X} - (\sigma/\sqrt{n})k_{\alpha/2},\ \bar{X} + (\sigma/\sqrt{n})k_{\alpha/2})$. Thus for any observed values $x_1, \ldots, x_n$ we have the interval estimate $(\bar{x} - (\sigma/\sqrt{n})k_{\alpha/2},\ \bar{x} + (\sigma/\sqrt{n})k_{\alpha/2})$. Moreover, by (2) we are assured that the interval so constructed will contain μ $100(1 - \alpha)$ percent of the time. Such interval estimates are commonly referred to as $100(1 - \alpha)$ percent confidence intervals. We shall adopt this terminology henceforth.

Example 2. The following data represent two samples of porosity measurements from two independent shipments of coke.

Sample 1 (X)

2.16	2.07	2.34	1.97
2.19	2.23	2.15	2.47
2.31	1.86	2.25	2.14
2.30	2.48	2.11	2.15
2.21	1.91	2.01	2.09

Sample 2 (Y)

1.97	1.90
2.31	1.94
2.15	2.16
2.24	2.04
2.07	2.25

$$n_X = 20 \qquad \sum_{i=1}^{20} X_i = 4340 \qquad \bar{X} = 2.170$$

$$\Sigma (X_i - \bar{X})^2 = .5146 \qquad s_X^2 = .0273,$$

$$n_Y = 10 \qquad \sum_{i=1}^{10} Y_i = 21.03 \qquad \bar{Y} = 2.103$$

$$\Sigma (Y_i - \bar{Y})^2 = .1444 \qquad s_Y^2 = .0144.$$

Find a 95 percent confidence interval on the true mean μ using the data from the first sample. Assume normality and σ^2 is known and is equal to $\frac{1}{16}$.

Solution: We know that

$$P\left[\bar{X} - k_{\alpha/2}\frac{\sigma}{\sqrt{n}} \leq \mu \leq \bar{X} + k_{\alpha/2}\frac{\sigma}{\sqrt{n}}\right] = 1 - \alpha.$$

For our problem, $1 - \alpha = .95$ or $\alpha = .05$,

$$\sigma = \sqrt{\sigma^2} = \tfrac{1}{4},$$

$$k_{\alpha/2} = 1.96 \qquad \text{(from Table I)},$$

$$\sqrt{n} = \sqrt{20} = 4.48,$$

$$\bar{X} = 2.170,$$

and hence $2.170 - 1.96\ (.250/4.48) \leq \mu \leq 2.170 + 1.96\ (.250/4.48)$ or $2.070 \leq \mu \leq 2.270$.

Example 3. Determine a $100(1 - \alpha)$ percent confidence interval for σ^2 in a $N(\mu, \sigma^2)$ distribution.

Solution: By Theorem 2, Section 4.9, we know that $nS^2/\sigma^2 \sim x^2(n - 1)$, where

$$S^2 = \frac{1}{n} \sum_{i=1}^{n} (X_i - \bar{X})^2.$$

But if χ_β^2 denotes a value which a chi-square random variable with $n - 1$ degrees of freedom exceeds only 100β percent of the time, we have

$$\int_{\chi^2_{1-\alpha/2}}^{\chi^2_{\alpha/2}} f_Y(y)\,dy = 1 - \alpha, \tag{3}$$

where $Y \sim \chi^2(n-1)$. Thus

$$P\left[\chi^2_{1-\alpha/2} \le \frac{nS^2}{\sigma^2} \le \chi^2_{\alpha/2}\right] = 1 - \alpha$$

or

$$P\left[\frac{nS^2}{\chi^2_{\alpha/2}} \le \sigma^2 \le \frac{nS^2}{\chi^2_{1-\alpha/2}}\right] = 1 - \alpha. \tag{4}$$

Equation (4) then defines the desired confidence interval.

Example 4. Find a 95 percent confidence interval for σ^2 using the data from the first sample in Example 1.

Solution: By (4)

$$P\left[\frac{\Sigma(X_i - \overline{X})^2}{\chi^2_{\alpha/2}} \le \sigma^2 \le \frac{\Sigma(X_i - \overline{X})^2}{\chi^2_{1-\alpha/2}}\right] = 1 - \alpha.$$

For our problem

$$\begin{aligned} \Sigma(x_i - \bar{x})^2 &= .5146, \\ n - 1 &= 19, \\ \alpha &= .05, \\ \alpha/2 &= .025, \\ 1 - \alpha/2 &= .975, \\ \chi^2_{\alpha/2} &= 32.9 \text{ (from Table II)}, \end{aligned}$$

and

$$\chi^2_{1-\alpha/2} = 8.91 \text{ (from Table II)}.$$

Hence

$$\frac{.5146}{32.9} \le \sigma^2 \le \frac{.5146}{8.91}$$

or

$$.015 \le \sigma^2 \le .577$$

is the desired 95 percent confidence interval for σ^2.

In Example 1, we found a confidence interval for μ in a normal distribution assuming σ was known. Practically speaking, σ is never known, although it may be essentially known when large sample sizes are involved. We can, however, escape this problem as follows. Recall from Chapter IV that for a random sample of size n from a normal distribution, if U and V are independent random variables such that

$$U \sim N(0, 1)$$

and

$$V \sim \chi^2(k),$$

then

$$\frac{U}{\sqrt{V}}\sqrt{k} \sim t(n - 1),$$

that is, has a t-distribution with $n - 1$ degrees of freedom. But

$$\frac{\bar{X} - \mu}{\sigma}\sqrt{n} \sim N(0, 1)$$

and

$$\frac{nS^2}{\sigma^2} \sim \chi^2(n - 1).$$

Moreover, they are both stochastically independent. Thus we have

$$\left(\frac{\bar{X} - \mu}{\sigma}\sqrt{n} \Big/ \sqrt{\frac{nS^2}{\sigma^2}}\right)\sqrt{n - 1} = \frac{\bar{X} - \mu}{S}\sqrt{n - 1} \sim t(n - 1). \tag{5}$$

Using the same notation as in the above examples, we can write

$$P[-t_{\alpha/2} \le T \le t_{\alpha/2}] = \int_{-t_{\alpha/2}}^{t_{\alpha/2}} f_T(t)\,dt = 1 - \alpha. \tag{6}$$

By (5) and (6),

$$P\left[-t_{\alpha/2} \le \frac{\bar{X} - \mu}{S}\sqrt{n - 1} \le t_{\alpha/2}\right] = 1 - \alpha,$$

or rearranging we have

$$P\left[\bar{X} - t_{\alpha/2}\frac{S}{\sqrt{n - 1}} \le \mu \le \bar{X} + t_{\alpha/2}\frac{S}{\sqrt{n - 1}}\right] = 1 - \alpha, \tag{7}$$

which defines a $100(1 - \alpha)$ percent confidence interval for μ when σ is unknown.

Example 5. Find a 95 percent confidence interval for μ when σ^2 is assumed unknown using the data from the first sample in Example 2.

Solution: By (7) such a confidence interval I is given by

$$I = \left(\bar{x} - t_{\alpha/2} \frac{s}{\sqrt{n - 1}}, \bar{x} + t_{\alpha/2} \frac{s}{\sqrt{n - 1}} \right).$$

But

$$\begin{aligned} \bar{x} &= 2.170, \\ \alpha/2 &= .025, \\ \sqrt{n - 1} &= 4.359, \\ s &= .165, \end{aligned}$$

and

$$t_{\alpha/2} = 2.093 \qquad \text{(from Table III)}.$$

Thus

$$I = (2.141, 2.199),$$

that is,

$$2.141 \leq \mu \leq 2.199$$

with 95 percent confidence.

Several interesting extensions to estimating the differences of two means are immediate. Suppose that there are two normal populations from which one can sample and we wish to discover whether or not their means are the same. Let

$$X \sim N_X(x; \mu_X, \sigma_X^2)$$

and

$$Y \sim N_Y(y; \mu_Y, \sigma_Y^2).$$

Also let $X_1, \ldots, X_n$ and $Y_1, \ldots, Y_m$ denote independent random samples from the two populations. Then

$$\bar{X} = \sum_{i=1}^{n} \frac{X_i}{n} \sim N\left(\mu_X, \frac{\sigma_X^2}{n}\right),$$

$$\bar{Y} = \sum_{i=1}^{m} \frac{Y_i}{n} \sim N\left(\mu_Y, \frac{\sigma_Y^2}{m}\right),$$

and

$$\overline{X} - \overline{Y} \sim N\left(\mu_X - \mu_Y, \frac{\sigma_X^2}{n} + \frac{\sigma_Y^2}{m}\right).$$

If σ_X^2 and σ_Y^2 are known, we can find a confidence interval for $\mu_X - \mu_Y$ as follows.

$$[\overline{X} - \overline{Y} - (\mu_X - \mu_Y)] \Big/ \left(\frac{\sigma_X^2}{n} + \frac{\sigma_Y^2}{m}\right)^{1/2} \sim N(0,1),$$

and the probability statement follows as before; that is,

$$P\left[-k_{\alpha/2} \leq \frac{\overline{X} - \overline{Y} - (\mu_X - \mu_Y)}{(\sigma_X^2/n + \sigma_Y^2/m)^{1/2}} \leq k_{\alpha/2}\right] = 1 - \alpha, \qquad (8)$$

which reduces to the confidence statement

$$P[\overline{X} - \overline{Y} - (\sigma_X^2/n + \sigma_Y^2/m)^{1/2} k_{\alpha/2} \leq \mu_X - \mu_Y$$
$$\leq \overline{X} - \overline{Y} + (\sigma_X^2/n + \sigma_Y^2/m)^{1/2} k_{\alpha/2}] = 1 - \alpha.$$

If σ_X^2 and σ_Y^2 are unknown, the fact that

$$\sum_{i=1}^{n} (X_i - \overline{X})^2/\sigma_X^2 \sim \chi^2(n - 1)$$

and

$$\sum_{i=1}^{m} (Y_i - \overline{Y})^2/\sigma_Y^2 \sim \chi^2(m - 1)$$

along with the assumption that $\sigma_X^2 = \sigma_Y^2 = \sigma^2$ gives easily a confidence bound for $\mu_X - \mu_Y$ using the t probability density. We observe that

$$\frac{\overline{X} - \overline{Y} - (\mu_X - \mu_Y)}{\sigma(1/n + 1/m)^{1/2}} \sim N(0,1)$$

and the sum of independent chi-square variates

$$\frac{\Sigma(X_i - \overline{X})^2}{\sigma^2} + \frac{\Sigma(Y_i - \overline{Y})^2}{\sigma^2} \sim \chi^2(n + m - 2)$$

lead naturally to a t-statistic void of the parameter σ^2. The confidence statement can be derived from the probability statement

$$P\left[-t_{\alpha/2}(n+m-2) \le \frac{\bar{X}-\bar{Y}-(\mu_X-\mu_Y)[mn/(m+n)]^{1/2}}{\left[\dfrac{\Sigma(X_i-\bar{X})^2}{n-1}+\dfrac{\Sigma(Y_i-\bar{Y})^2}{m-1}\right]^{2}} \le t_{\alpha/2}(n+m-2)\right] = 1-\alpha,$$

by simple manipulation. The result is

$$P\left[\bar{X}-\bar{Y}-\left(\frac{m+n}{mn}\right)^{1/2}\left[\frac{\Sigma(X_i-\bar{X})^2}{n-1}+\frac{\Sigma(Y_i-\bar{Y})^2}{m-1}\right]^{1/2} t_{\alpha/2}(n+m-2) \le \mu_X-\mu_Y \le \bar{X}-\bar{Y}+\left(\frac{m+n}{mn}\right)^{1/2}\left[\frac{\Sigma(X_i-\bar{X})^2}{n-1}+\frac{\Sigma(Y_i-\bar{Y})^2}{m-1}\right]^{1/2} t_{\alpha/2}(n+m-2)\right] = 1-\alpha. \quad (9)$$

Approximate confidence intervals for $\mu_X - \mu_Y$ exist when $\sigma_X^2 \neq \sigma_Y^2$, but they will not be discussed here.

Example 6. Do the data in Samples 1 and 2 in Example 1 justify the conclusion that the mean porosity μ_X is different from the mean porosity μ_Y? Select $\alpha = .05$.

Solution: The question asked is simply: Is $\mu_X \neq \mu_Y$, that is, is $\mu_X - \mu_Y \neq 0$? We proceed by estimating the difference by a confidence interval estimate. If σ_X^2 and σ_Y^2 are known, then (8) applies. If $\sigma_X^2 = \sigma_Y^2 = \sigma^2$, where σ^2 is unknown, then (9) applies.

(a) Let $\sigma_X^2 = \frac{1}{16}$ and $\sigma_Y^2 = \frac{1}{25}$; then

$$\bar{x}-\bar{y} = 2.170 - 2.103 = .067,$$

$$k_{\alpha/2} = k_{.025} = 1.96,$$

$$n = 20,$$

$$m = 10,$$

$$\bar{x} - \bar{y} - k_{\alpha/2}\sqrt{\frac{\sigma_X^2}{n} + \frac{\sigma_Y^2}{m}} = .067 - 1.96\sqrt{\frac{1}{320} + \frac{1}{250}}$$

$$= -.097,$$

and

$$\bar{x} + \bar{y} + k_{\alpha/2}\sqrt{\frac{\sigma_X^2}{n} + \frac{\sigma_Y^2}{m}} = .067 + 1.96\sqrt{\frac{1}{320} + \frac{1}{250}}$$

$$= .231.$$

Hence the 95 percent confidence interval estimate for $\mu_X - \mu_Y$ is

$$-.097 \leq \mu_X - \mu_Y \leq .231.$$

We see that zero is in the interval, so there is no reason statistically, at least, for us to believe that

$$\mu_X \neq \mu_Y.$$

(b) If σ_X^2 and σ_Y^2 are unknown, but it can reasonably be assumed that

$$\sigma_X^2 = \sigma_Y^2 = \sigma^2,$$

then we apply (9). We compute

$$\bar{x} - \bar{y} = .067,$$

$$n + m - 2 = 20 + 10 - 2 = 18 \text{ degrees of freedom},$$

$$t_{.025}(18) = 2.101 \quad \text{(from Table III)},$$

$$\left(\frac{m + n}{mn}\right)^{1/2} = \left(\frac{30}{200}\right)^{1/2} = (.15)^{1/2} = .39,$$

and

$$\left[\frac{\Sigma(X_i - \bar{X})^2}{n - 1} + \frac{\Sigma(Y_i - \bar{Y})^2}{m - 1}\right]^{1/2} = .28.$$

Hence the 95 percent confidence interval estimate for $\mu_X - \mu_Y$ is

$$.067 - 2.101(.39)(.28) \leq \mu_X - \mu_Y \leq .067 + 2.101(.39)(.28),$$

$$-.161 \leq \mu_X - \mu_Y \leq .295.$$

Again, we note that zero is contained in the interval, and hence we conclude that the data do not support the conjecture that $\mu_X \neq \mu_Y$.

A somewhat different approach is necessary to compare variance, say σ_X^2 and σ_Y^2, of two normal populations. In Chapter IV it was noted that the ratio of two independent chi-square random variables each divided by their respective degrees of freedoms, m and n, is distributed according to an F probability density function with parameters m and n.

We recall that

$$\frac{\Sigma(X_i - \bar{X})^2}{\sigma_X^2} \sim \chi^2(n - 1)$$

and

$$\frac{\Sigma(Y_i - \bar{Y})^2}{\sigma_Y^2} \sim \chi^2(m - 1).$$

Hence

$$Z = \frac{\dfrac{\Sigma(X_i - \bar{X})^2}{n - 1}}{\dfrac{\Sigma(Y_i - \bar{Y})^2}{m - 1}} \frac{\sigma_Y^2}{\sigma_X^2}$$

is distributed as an F density with degrees of freedom $n - 1$ and $m - 1$; that is, $Z \sim F(n - 1, m - 1)$. Thus

$$P\left[Z_{1-\alpha/2}(n - 1, m - 1) \leq \frac{\dfrac{\Sigma(X_i - \bar{X})^2}{n - 1}}{\dfrac{\Sigma(Y_i - \bar{Y})^2}{m - 1}} \frac{\sigma_Y^2}{\sigma_X^2} \leq Z_{\alpha/2}(n - 1, m - 1)\right] = 1 - \alpha, \quad (10)$$

where

$$\int_{Z_\beta}^{\infty} f_Z(z; n - 1, m - 1)\, dz = \beta,$$

and $f_Z(z;\, n - 1,\, m - 1)$ denotes the F probability density function with $n - 1$ and $m - 1$ degrees of freedom. The values for Z_β for various values of n, m, and β are given in Table IV. By simple manipulation, we can write (10) in the form

$$P\left[L \leq \frac{\sigma_X^2}{\sigma_Y^2} \leq U\right] = 1 - \alpha, \tag{11}$$

where

$$L = \frac{\Sigma(Y_i - \bar{Y})^2/m - 1}{\Sigma(X_i - \bar{X})^2/n - 1} Z_{1-\alpha/2}$$

and

$$U = \frac{\Sigma(Y_i - \bar{Y})^2/m - 1}{\Sigma(X_i - \bar{X})^2/n - 1} Z_{\alpha/2}.$$

A statistical decision can be made concerning the equality of the variances σ_X^2 and σ_Y^2 by noting whether or not the interval (L, U) contains the value unity, since $\sigma_X^2 = \sigma_Y^2$ if and only if

$$\frac{\sigma_X^2}{\sigma_Y^2} = 1.$$

Example 7. Is there information in the two samples of Example 1 to justify the conclusion that

$$\sigma_X^2 = \sigma_Y^2$$

at the 95 percent confidence level? Let $\alpha = 0.05$.

Solution: Let us estimate the ratio σ_X^2/σ_Y^2 by a confidence interval estimator using (11). We compute

$$Z_{.025}(19,9) = 4.81 \qquad \text{(from Table IV)},$$

$$Z_{.975}(19,9) = .25 \qquad \text{(from Table IV)},$$

$$L = \frac{.0160}{.0271}.25 = .148,$$

$$U = \frac{.0160}{.0271}4.81 = 2.838.$$

The 95 percent confidence interval estimate of the ratio σ_X^2/σ_Y^2 is, therefore,

$$148 \leq \frac{\sigma_X^2}{\sigma_Y^2} \leq 2.838.$$

We note that unity is in the interval; hence the data do support, at this level of confidence, the conjecture that $\sigma_X^2 = \sigma_Y^2$.

We freely admit that we have not strayed far from the assumption that the observations are normal. Suppose, for demonstration purposes, that

$$f_X(x;\theta) = \begin{cases} \frac{1}{\theta} e^{-(1/\theta)x}, & 0 < x, \theta > 0, \\ 0, & \text{elsewhere,} \end{cases}$$

and $X_1, X_2, \ldots, X_n$ devotes a random sample from this density. Our project is to formulate a confidence interval estimator for θ.

The likelihood function is

$$L = L(x_1, \ldots, x_n) = \frac{1}{\theta^n} e^{-(1/\theta)\Sigma x_i}.$$

It then follows that

$$\ln L = -n \ln \theta - \frac{1}{\theta} \Sigma x_i.$$

A necessary condition for L to be maximum with respect to θ is that

$$\frac{d \ln L}{d\theta} = -\frac{n}{\theta} + \frac{1}{\theta^2} \Sigma x_i = 0.$$

This equation yields the maximum likelihood estimate

$$\hat{\theta} = \sum_{i=1}^{n} x_i / n.$$

But the moment-generating function for X is simply

$$M_X(t) = \frac{1}{1 - \theta t};$$

hence

$$M_{\Sigma X_i}(t) = \frac{1}{(1 - \theta t)^n}$$

and

$$M_{(2/\theta)\Sigma X_i}(t) = \frac{1}{(1 - 2t)^n} = \frac{1}{(1 - 2t)^{2n/2}}.$$

Our manipulations have produced a moment-generating function for a chi-square variate with $2n$ degrees of freedom, that is,

$$\frac{2}{\theta} \Sigma X_i \sim \chi^2(2n),$$

which leads to the probability statement

$$P\left[\chi^2_{1-\alpha/2}(2n) \leq \frac{2\Sigma X_i}{\theta} \leq \chi^2_{\alpha/2}(2n)\right] = 1 - \alpha.$$

This, in turn, defines an interval estimator for θ by means of the equation

$$P\left[\frac{2\sum_{i=1}^{n} X_i}{\chi^2_{\alpha/2}(2n)} \leq \theta \leq \frac{2\sum_{i=1}^{n} X_i}{\chi^2_{1-\alpha/2}(2n)}\right] = 1 - \alpha.$$

The above illustrates roughly a procedure one often follows to obtain interval estimators for parameters when the assumption of normality is unsatisfactory.

VI
Stochastic Processes

6.1 INTRODUCTION

In recent years an area of probability referred to as stochastic processes has become increasingly popular and useful in the fields of engineering and applied mathematics. In this chapter we introduce briefly some of the more important concepts and results. A thorough study of the subject involves quite sophisticated mathematics and hence is not our purpose. We will, however, try to develop some feel for the subject and its applications. To do this it will be necessary to introduce some additional definitions which we will set apart for emphasis.

Definition 1 Let T be any set and let $X(t)$ be a random variable for each $t \in T$. (The notation $t \in T$ will be used extensively in this chapter. It means that t is some real number in the set of real numbers T.) Then we define a stochastic process as the collection of random variables $\{X(t); t \in T\}$.

In Chapter II we discussed the notion of a random variable. Roughly, we saw that a random variable is simply a particular type of function. By introducing this concept we found that the wide diversity of probability statements and questions regarding physical phenomena could be analyzed by associating a probability structure with the manner in which a random variable assumes its values. Thus in solving the problem (Example 1, Section 4.6) of determining the probability of no failures of a certain radar in a 400-hour period, we let X be a function (random variable) which assumed its values according to the probability law

$$P[X = x] = \begin{cases} \dfrac{2^x e^{-2}}{x!}, & x = 0,1,2,\ldots, \\ 0, & \text{elsewhere.} \end{cases}$$

Now according to Definition 1 a stochastic process $\{X(t); t \in T\}$ is a collection of random variables. The above discussion reminds us that this is not a collection of real numbers but a collection of functions. Thus, if $t = t_0 \in T$, $X(t_0)$ is not a number but is the particular random variable associated with the index t_0. The values of $X(t_0)$ are, of course, real numbers and $X(t_0)$ assumes these values according to some probability law. From this point of view the notation $\{X_t; t \in T\}$ might be less confusing, but this would be an inconvenient and a rather foreign notation in developments which follow. As in the previous chapters, we will denote random variables by capital letters and their values by lowercase letters. Thus $X(t)$ is a random variable, while $x(t)$ is a value of $X(t)$. We will generally refer to $x(t)$ as a sample function and the set $\{x_k(t); t \in T; k = 1,2,\ldots,n\}$ as an ensemble of sample functions. Consider now the following examples.

Example 1. Let $T = \{2,3,\ldots,n\}$ and $\{X(t), t \in T\}$ be a collection of independent random variables such that

$$P[X(t) = k] = \begin{cases} \dbinom{m}{k}\left(\dfrac{1}{t}\right)^k \left(1 - \dfrac{1}{t}\right)^{m-k}, & k = 0,1,\ldots,m, \\ 0, & \text{elsewhere.} \end{cases} \tag{1}$$

Then $\{X(t), t \in T\}$ is a stochastic process.

From Table 4.1 we note that in this case for each fixed t

$$E[X(t)] = \frac{m}{t}$$

and

$$\operatorname{Var} X(t) = \frac{m(t-1)}{t^2}.$$

For a sample of size one from each of the $X(t)$ populations we have a typical plot of the observation $x(t)$ shown in Figure 6.1.

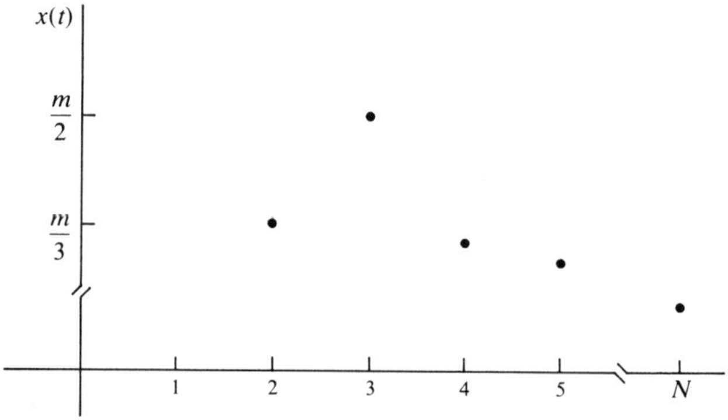

Figure 6.1 Binomial Realization

To interpret Figure 6.1, note that equation (1) defines a probability model for $X(t)$ for each t. Thus for $t = 2$,

$$P[X(2) = k] = \binom{m}{k}\left(\frac{1}{2}\right)^k\left(\frac{1}{2}\right)^{m-k},$$

while, if $t = 3$,

$$P[X(3) = k] = \binom{m}{k}\left(\frac{1}{3}\right)^k\left(\frac{2}{3}\right)^{m-k}.$$

Figure 6.1 then represents a sequence of outcomes or observations, that is, one from each of the distributions defined by equation (1).

Example 2. Let $T = \{t;\ -\infty < t < \infty\}$ and $X(t) = \sin(\omega t + \Phi)$, where $\Phi \sim N(0,1)$. Then $\{X(t);\ t \in T\}$ is a stochastic process. Typical sample functions, that is, typical values of $X(t)$ for this process, are shown in Figure 6.2. It should be emphasized that Figure 6.2 presents graphs of values of $X(t)$, that is, graphs of $x(t) = \sin(\omega t + \phi)$, where ϕ is a particular value of Φ. Indeed, it is impossible to plot $X(t)$ since its graph as a functon of t is different for different values of Φ.

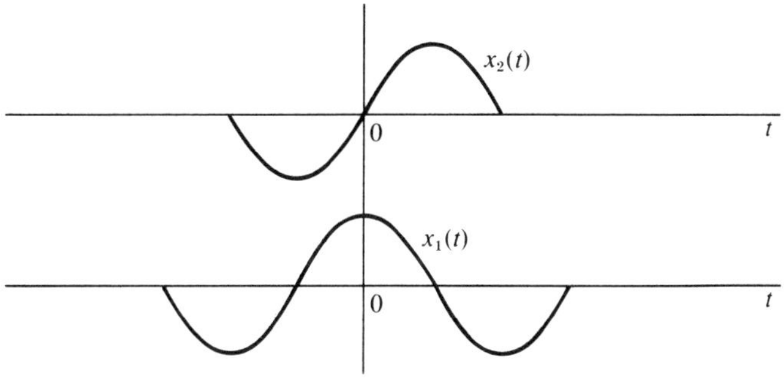

Figure 6.2 $X(t) = \sin(\omega t + \phi)$

Example 3. Let $T = \{1,2,3,\ldots,365\}$. Also let $X(t)$ be a random variable which maps the wind velocity at Lubbock, Texas, on the tth day of the year onto the real number k, where k is the numerical value of the average wind velocity in miles per hour on the tth day. Thus, if the average wind velocity on January 1, at Lubbock, is 35 mph, we have $x(1) = 35$. Clearly, $\{X(t), t \in T\}$ could be considered a stochastic process.

Example 4. Suppose the output voltage of a thermal noise generator is recorded as a function of time. A specific voltage-time history will then be obtained. However, if a second time

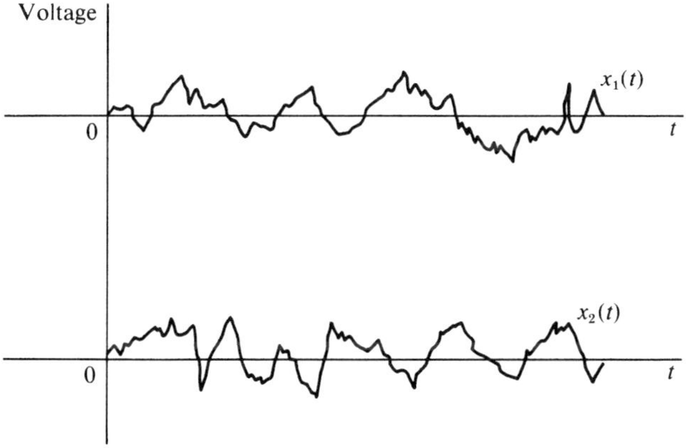

Figure 6.3 Thermal Noise

history is simultaneously recorded from an "identical" generator, its time history would be different, as shown in Figure 6.3.

Clearly at each t, one sees that the voltage takes on values which might best be described through the laws of probability. Hence if we let $x(t)$ = numerical value of the voltage at time t, we can interpret Figure 6.3 as observations (or sample functions) from a stochastic process $\{X(t);\ t \in T\}$, where $T = \{t; 0 \leq t < \infty\}$.

Example 5. Let $X(t) = S(t) + N(t)$, where $S(t)$ is the true velocity of a satellite at time t and $N(t)$ is a random-noise variable, that is, measurement errors. Clearly, $\{X(t); 0 < t < \infty\}$ is a stochastic process. In this case one observes $X(t)$ and attempts to draw conclusions regarding $S(t)$. Thus one may wish to estimate $S(t_0)$, or predict $S(t_0 + \tau)$, $\tau > 0$, by observing $X(t)$ for $t \leq t_0$. These latter two problems are referred to as the *prediction*, and estimation or filter, *problem*.

Example 6. Physical characteristics, such as weight, density, etc., of an object or particle, often vary with time in such a manner that these characteristics can best be described by means of a set random variable with time as an index set. Thus such things as (a) the population explosion, (b) radioactive decay, (c) atomic fission, (d) growth of a fungus, etc., can be considered as stochastic processes. In fact, any process which evolves with time according to probabilistic laws can be interpreted as a stochastic process where each time history record is considered as a sample function.

In the previous examples we have referred to a sample function as an observation. We will often do this, although it should be pointed out again that a sample function is actually a collection of observations, one from each of the random variables $X(t)$. To make this distinction we will use the word "realization" when it appears that the word "observation" may be confusing.

6.2 THE PROBABILITY DENSITY FUNCTION AND MOMENTS

We have already discussed probability density functions and moments (Chapters II, III, and IV). However, since some people find the index t confusing and since these quantities are of primary importance, we shall discuss the probability density and first- and second-order moments once more.

Since a stochastic process $\{X(t);\ t \in T\}$ is a collection of random variables, there is a probability density function associated with each t. This is, however, easily accounted for in our current notation. That is, there should be no confusion when we use the notation $f_{X(t_0)}$ for the pdf associated with the random variable $X(t)$ when $t = t_0$. Of course, in some cases $f_{X(t)}$ will be the same pdf for each t and in that case we will drop the t and simply write f_X.

Definition 1 Let $\{X(t);\ t \in T\}$ be stochastic process. Then for each fixed t the $E[X(t)]$ is given by

$$E[X(t)] = \begin{cases} \int_{-\infty}^{\infty} x(t) f_{X(t)}[x(t)]\, dx(t), & X(t) \text{ continuous}, \\ \sum_{k=0}^{\infty} x_k(t) f_{X(t)}[x_k(t)], & X(t) \text{ discrete}. \end{cases} \tag{1}$$

We should elaborate some on Definition 1. First, it should be noted that it is precisely the same as Definition 1 in Chapter III. This is clear if one remembers that for a fixed t, $X(t)$ is simply a random variable and that the letter t is simply playing the role of an index. Thus for, say $t = t_0$, if we let $Y = X(t_0)$ we see by (1) that

$$E[X(t_0)] = E[Y] = \int_{-\infty}^{\infty} y f_Y(y)\, dy, \tag{2}$$

if Y is a continuous random variable. From the above it is clear that $E[X(t)]$ may be a different number for different values of t. Thus we will often use the notation $\mu(t)$ for $E[X(t)]$; that is, $E[X(t)] = \mu(t)$.

Definition 2 Let $\{X(t);\ t \in T\}$ be a stochastic process with $X(t)$ a continuous random variable for each $t \in T$.

Then for any t_0 and $t_1 \in T$ we define

(a) the autocorrelation $R_X(t_0, t_1)$ by

$$R_X(t_0, t_1) = E[X(t_0)X(t_1)]$$
$$= \int_{-\infty}^{\infty}\int_{-\infty}^{\infty} x(t_0)x(t_1)f_{X(t_0)X(t_1)}$$
$$[x(t_0), x(t_1)]\,dx(t_0)\,dx(t_1) \qquad (3)$$

and

(b) the autocovariance $C_X(t_0, t_1)$ by

$$C_X(t_0, t_1) = E[(X(t_0) - \mu(t_0))(X(t_1) - \mu(t_1))]$$
$$= \int_{-\infty}^{\infty}\int_{-\infty}^{\infty} [x(t_0) - \mu(t_0)][x(t_1) - \mu(t_1)]f_{X(t_0)X(t_1)}$$
$$(x(t_0), x(t_1))\,dx(t_0)\,dx(t_1)$$
$$= R_X(t_0, t_1) - \mu(t_0)\mu(t_1), \qquad (4)$$

where $f_{X(t_0)X(t_1)}$ is the joint pdf for $X(t_0)$ and $X(t_1)$.

We again remark that equations (3) and (4) are no different from our previous definitions. Thus, if $X(t_0) = Y$ and $X(t_1) = Z$, we have for equation (3)

$$E[X(t_0)X(t_1)] = E[YZ]$$
$$= \int_{-\infty}^{\infty}\int_{-\infty}^{\infty} yzf_{YZ}(y, z)\,dy\,dz,$$

which is the same as previously given. One should note in (4) that if $t_0 = t_1$ we have

$$C_X(t_0, t_1) = C_X(t_0, t_0)$$
$$= E[(X(t_0) - \mu(t_0))^2]$$
$$= \sigma_X^2(t_0)$$
$$= R_X(t_0, t_0) - [\mu(t_0)]^2.$$

The condition in Definition 2 that $X(t)$ is continuous can be replaced by the condition that $X(t)$ is discrete by replacing integrals by sums.

Definition 3 Let $\{X(t);\ t \in T\}$ and $\{Y(t);\ t \in T\}$ be stochastic processes and t_0 and t_1 be any two points in T. Then the cross correlation $R_{XY}(t_0, t_1)$ and cross covariance $C_{XY}(t_0, t_1)$ are defined by

$$R_{XY}(t_0, t_1) = E[X(t_0)Y(t_1)], \tag{5}$$

$$\begin{aligned} C_{XY}(t_0, t_1) &= E[(X(t_0) - \mu_x(t_0))(Y(t_1) - \mu_y(t_1))] \\ &= R_{XY}(t_0, t_1) - \mu_x(t_0)\mu_y(t_1). \end{aligned} \tag{6}$$

Example 1. In Example 1, Section 6.1,

$$f_{X(t)}(x(t)) = \begin{cases} \binom{m}{x(t)}\left(\frac{1}{t}\right)^{x(t)}\left(1 - \frac{1}{t}\right)^{m-x(t)}, & x(t) = 0,1,\ldots,m \\ 0, & \text{elsewhere.} \end{cases} \tag{7}$$

Thus the range of $x(t)$ is independent of t; that is, $x(t) = 0,1,\ldots,m$ when $f_{X(t)}(x(t)) \neq 0$ for every t. When this is the case the t in the notation $x(t)$ conveys no information and we will therefore delete it. Thus we write

$$f_{X(t)}(x) = \begin{cases} \binom{m}{x}\left(\frac{1}{t}\right)^{x}\left(1 - \frac{1}{t}\right)^{m-x}, & x = 0, 1, \ldots, m \\ 0, & \text{elsewhere} \end{cases} \tag{8}$$

rather than equation (7).

Clearly,

$$\mu(t) = \frac{m}{t}.$$

Now suppose $X(t_0)$ and $X(t_1)$ are independent for any t_0 and t_1. Then

$$\begin{aligned} R_X(t_0, t_1) &= E[X(t_0)X(t_1)] \\ &= E[X(t_0)]E[X(t_1)] \\ &= \frac{m}{t_0}\frac{m}{t_1} \\ &= \frac{m^2}{t_0 t_1} \end{aligned}$$

and

$$C_X(t_0, t_1) = \frac{m^2}{t_0 t_1} - \frac{m}{t_0}\frac{m}{t_1} = 0. \tag{9}$$

Example 2. Let $X(t) = \sin(\omega t + \Phi)$, where $\Phi \sim U(-t, t)$. Then

$$\begin{aligned} E[X(t)] &= \frac{1}{2t}\int_{-t}^{t} \sin(\omega t + \phi)\, d\phi \\ &= \frac{1}{2t}[\cos(\omega t - t) - \cos(\omega t + t)] \\ &= \frac{1}{t}\sin \omega t \sin t. \end{aligned}$$

If $X(t_0)$ and $X(t_1)$ are independent, then

$$\begin{aligned} R_X(t_0, t_1) &= E[X(t_0)]E[X(t_1)] \\ &= \frac{1}{t_0 t_1} \sin \omega t_0 \sin \omega t_1 \sin t_1, \end{aligned}$$

$$C_X(t_0, t_1) = 0.$$

6.3 TIME AVERAGES

In Section 6.2 we discussed the first- and second-order moments associated with a stochastic process. In Chapter V we saw that the sample moments were useful as estimators for these population parameters. In attempting to make use of this fact in estimating the parameters of a stochastic process, we are faced with the difficulty that often only parts of one or two realizations are available for analysis.

This means that we may be attempting to estimate the probability structure of an uncountable number of distinct random variables from only one or two observations of each. Thus if only one realization is available and we wish to estimate $E[X(t_0)]$, then $\bar{x}(t_0) = x(t_0)$, since this is the only observation we have. From this viewpoint the problem of estimating quantities such as $\mu(t)$, $R_X(t_0, t_1)$, etc., to any reasonable degree of accuracy seems insurmountable. However, in many instances of practical interest this is not the case. In fact, when one can justify an appropriate ergodicity assumption these parameters can be estimated to any desired degree of accuracy from this seemingly sparse amount of data. To clarify this consider Figure 6.4. From the nature of the graphs in this figure it seems likely that for any given t if we averaged $x(t)$ over a large number of realizations the result should be independent of t. That is, suppose we have n realizations, $x_1(t)$, $x_2(t), \ldots,$ $x_n(t)$ and t_0 and t_1 are arbitrary values of t.

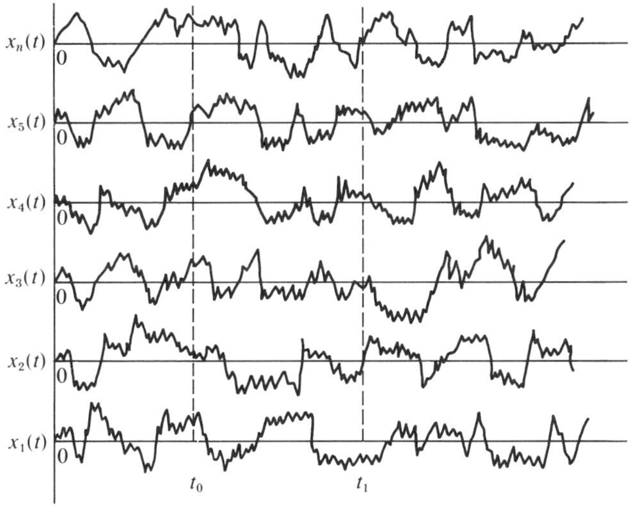

Figure 6.4 "Ergodic" Noise

Then if

$$\bar{x}_n(t_0) = \frac{1}{n}\sum_{i=1}^{n} x_i(t_0) \tag{1}$$

and

$$\bar{x}_n(t_1) = \frac{1}{n}\sum_{i=1}^{n} x_i(t_1), \tag{2}$$

we might expect $\bar{x}_n(t_0) \approx x_n(t_1)$ for large n. In fact, it is not unreasonable to claim that

$$\lim_{n\to\infty} \bar{x}_n(t_0) = \lim_{n\to\infty} \bar{x}_n(t_1) = E[X(t)] = \mu,$$

since, by considering the random behavior of the graphs in Figure 6.4, both across the ensemble and down the time axis, we might also expect that

$$\frac{1}{n}\sum_{t=0}^{n-1} x_k(t) \approx \bar{x}_n(t_0) \approx \bar{x}_n(t_1) \qquad \text{for any } k \tag{3}$$

and

$$\lim_{n\to\infty} \frac{1}{n}\sum_{t=0}^{n-1} x_k(t) = \lim_{n\to\infty} \bar{x}_n(t_0) = \lim_{n\to\infty} \bar{x}_n(t_1) = E[X(t)]. \tag{4}$$

Note that on the left side of (3) we are averaging along the time axis, while on the right side we are averaging across the ensemble.

From Chapter III we know, speaking rather loosely, that $\bar{x}_n(t_0) \to E[X(t_0)]$ as $n \to \infty$ and $\bar{x}_n(t_1) \to E[X(t_1)]$ as $n \to \infty$. We see for a stochastic process such as the one we are discussing that we have $E[X(t_0)] = E[X(t_1)]$ for any t_0 and t_1 and hence $E[X(t)]$ must be independent of t, that is, $E[X(t)] = \mu$, a constant.

The above discussion suggests two important concepts. First, we see that in some cases it is reasonable for the $E[X(t)]$ to be independent of t and, second, it may be possible to estimate this quantity by averaging along the time axis rather than across the ensemble. When the latter is true we call the process ergodic in the mean. For stochastic processes in which these ideas are plausible, interpretation of the probability structure of the process from only one or two realizations becomes more feasible. This is, of course, to be expected since we are then able to supplement the scarcity of "across-ensemble data" by the more plentiful data along the time axis. To solidify these thoughts we introduce the following definitions.

Definition 1 An index set T is said to be a linear index set if for any t_1 and $t_2 \in T$ we have $t_1 + t_2 \in T$.

Definition 2 A stochastic process $\{X(t);\ t \in T\}$ is said to be strictly stationary if

(a) T is a linear index set

and

(b) for any positive integer k and any $\tau \in T$ the k-dimensional random variables $(X(t_1), \ldots, X(t_k))$ and $(X(t_1 + \tau), \ldots, X(t_k + \tau))$ are identically distributed.

The requirement of strict stationarity is a severe one and one which is often very difficult to establish mathematically. However, there is a less restrictive requirement, known as weak stationarity, which often suffices in practical applications.

Definition 3 If T is a linear index set and if for every t and $\tau \in T$, $\mu(t)$ and $R_X(t, t + \tau)$ are independent of t, then we say that the stochastic process $\{X(t);\ t \in T\}$ is a weakly stationary process.

Thus a stochastic process is called weakly stationary if $E[X(t)]$ is independent of t and if for any t_0 and $t_1 \in T$ the autocorrelation $R_X(t_0, t_1)$ is a function only of $t_0 - t_1$. Unless stated to the contrary, any future reference to the word stationary will mean weakly stationary. Since in this case $R_X(t_0, t_1)$ is a function only of $\tau = t_1 - t_0$, we will adopt the shorter notation $R_X(\tau)$ for stationary processes. Similar alterations will be made in the notation of $R_{XY}(t_0, t_1)$ and $C_{XY}(t_0, t_1)$.

Definition 4 Let $\{X(t);\ t \in T\}$ and $\{Y(t);\ t \in T\}$ be stochastic processes. If T is a linear index set and $E[X(t)]$, $E[Y(t)]$, $R_X(t + \tau, t)$, $R_Y(t + \tau, t)$, and $R_{XY}(t + \tau, t)$ are finite and independent of t for every t and $\tau \in T$, then we say the two stochastic processes are jointly weakly stationary. Note that $R_{XY}(\tau) = E[X(t + \tau)Y(t)] = E[Y(t)X(t + \tau)] = R_{YX}(-\tau)$.

The extension of the idea of weak stationarity to any number of stochastic processes is clear. It is also now clear that when attempting to estimate ensemble moments such as μ and $R_X(\tau)$ by a sample time average we must restrict ourselves to processes which are stationary with regard to these moments. If a process is stationary with regard to a certain ensemble moment and if the corresponding sample time average converges to this parameter, then we say the process is ergodic with respect to that moment. Thus if $\{X(t);\ t \in T\}$ is stationary with respect to the mean and

$$\overline{X}(N) = \frac{1}{N} \sum_{t=0}^{N-1} X(t) \rightarrow \mu$$

as $N \rightarrow \infty$, we say the process is ergodic in the mean.

Immediately above we described an ergodic process as one in which ensemble averages could be replaced by time sample averages. We will now clarify the meaning of the words "time sample averages." Since from this point on all sample averages will be time averages, we will often delete the word "time" and simply say "sample average." Suppose we first consider the stationary stochastic process $\{X(t);\ t = 0,1,2,\ldots\}$. We define the sample mean $\overline{X}(N)$, the sample autocorrelation $\overline{R}_X(\tau)$, and sample covariance $\overline{C}_X(\tau)$ by

$$\overline{X}(N) = \frac{1}{N}\sum_{t=0}^{N-1} X(t), \tag{5}$$

$$\overline{R}_X(N;\tau) = \frac{1}{N}\sum_{t=0}^{N-1} X(t)X(t+\tau), \tag{6}$$

$$\overline{C}_X(N;\tau) = \overline{R}_X(N;\tau) - [\overline{X}(N)]^2. \tag{7}$$

For the stochastic processes $\{X(t);\ t = 0,1,\ldots\}$ and $\{Y(t);\ t = 0,1,\ldots\}$ the sample cross-correlation $\overline{R}_{XY}(N,\tau)$ is defined by

$$\overline{R}_{XY}(N,\tau) = \frac{1}{N}\sum_{t=0}^{N-1} X(t+\tau)Y(t). \tag{8}$$

We now extend these definitions in the obvious manner to the index set $T = \{t;\ 0 \le t < \infty\}$. If $T = \{t;\ 0 \le t < \infty\}$ we define the sample mean $\overline{X}(N)$, the sample autocorrelation $\overline{R}_X(N;\tau)$, and the sample covariance by

$$\overline{X}(N) = \frac{1}{N}\int_0^N X(t)\,dt, \tag{9}$$

$$\overline{R}_X(N;\tau) = \frac{1}{N}\int_0^N X(t)X(t+\tau)\,dt, \tag{10}$$

$$\overline{C}_X(N;\tau) = \overline{R}_X(N;\tau) - [\overline{X}(N)]^2. \tag{11}$$

Also, we define the sample cross correlation by

$$\overline{R}_{XY}(N;\tau) = \frac{1}{N}\int_0^N X(t+\tau)Y(t)\,dt.$$

The astute reader may wonder exactly what meaning should be attached to the integrals in (9), (10), and (11). We do not wish to expound on this for fear of creating confusion. We simply remark that the integrals here are limits in the mean-square sense. They are legitimate random variables and we will assume that for any random variables considered the property

$$E\left[\int_a^b X(t)\,dt\right] = \int_a^b E[X(t)]\,dt$$

holds. Similar comments are valid for the random variable $dX(t)/dt$. We will formulate most future results in such a manner that the reader who fails to see the point of this short discussion will not be seriously hampered. For further discussion see Papoulis (Ref. 17) or Parzen (Ref. 19).

6.4 ERGODICITY

As already mentioned, in many physical problems it may not be possible to obtain more than a portion of one or two sample functions. For this reason it is often impossible to do any significant data analysis unless one can justify an ergodicity assumption, since without this assumption one has almost no data. Hence the question of ergodicity is of paramount importance. For this reason we will now discuss some results which can be used to justify the assumption.

In the next theorem we will be interested in the quantity $\lim_{N\to\infty} \overline{X}(N)$. Just as in the case of the integral of a random variable there certainly is some question of what this limit means, since it is not the limit of a sequence of real numbers, but the limit of a sequence of functions. Our interpretation of the statement

$$\lim_{N\to\infty} \overline{X}(N) = \mu \tag{1}$$

will simply mean

$$\lim_{N\to\infty} \operatorname{Var} \overline{X}(N) = 0. \tag{2}$$

This is reasonable, since the variance of $\overline{X}(N)$ is a measure of its variation about μ. Thus (1) holds if the variation about μ

is zero in the limit. No explanation of (2) is necessary, since it is simply a limit of a sequence of real numbers. The definition is quite intuitive. It simply says if the $\overline{X}(N)$ converges to the constant μ, then its values must become increasingly concentrated about μ as N gets large.

Theorem 1 Let $\{X(t); t \in T\}$ be a weakly stationary process with finite mean and covariance, μ and $C_X(\tau)$, respectively. Then

(a) if $T = \{t; t = 0, 1, \ldots\}$ we have

$$\lim_{N\to\infty} \overline{X}(N) = \mu \tag{3}$$

if and only if

$$\lim_{N\to\infty} \frac{1}{N} \sum_{t=0}^{N-1} C_X(t) = 0, \tag{4}$$

and

(b) if $T = \{t; 0 \leq t < \infty\}$ we have

$$\lim_{N\to\infty} \overline{X}(N) = \mu \tag{5}$$

if and only if

$$\lim_{N\to\infty} \frac{1}{N} \int_0^N C_X(\tau)\, d\tau = 0. \tag{6}$$

Theorem 1 has a very simple, useful interpretation. First we point out that one can easily show that if $T = \{t; t = 0,1, \ldots\}$, then

$$\frac{1}{N} \sum_{t=0}^{N-1} C_X(t) = \operatorname{Cov}[\overline{X}(N), X(N-1)] \tag{7}$$

and if $T = \{t; 0 \leq t < \infty\}$, then

$$\frac{1}{N} \int_0^N C_X(\tau)\, d\tau = \operatorname{Cov}[\overline{X}(N), X(N)]. \tag{8}$$

Hence in either case the theorem simply tells us that a necessary and sufficient condition that the process be ergodic in the mean is that the sample mean and $X(N)$ become essentially uncorrelated for large values of N. Thus, as is reasonable, in a process which is ergodic in the mean the sample mean is not affected much by $X(N)$ if N is large.

Corollary 1 For the processes described in Theorem 1,

$$\lim_{N \to \infty} \overline{X}(N) = \mu \tag{9}$$

if

$$\lim_{\tau \to \infty} C_X(\tau) = 0. \tag{10}$$

By Corollary 1 any process in which $X(t)$ and $X(t + \tau)$ become uncorrelated as τ becomes large (that is, if the random variables exhibit only local correlation) is ergodic in the mean.

The previous theorem establishes a necessary and sufficient condition for the sample average $\overline{X}(N)$ to converge to μ. Now $\overline{X}(N)$ is a random variable and in practice we work with particular values of $\overline{X}(N)$. Thus from a practicing engineer's viewpoint the real question is what does the above theorem mean in terms of the observations $x(t)$. This is the context of the next theorem.

Theorem 2 Let $\{X(t)\}$ be a stationary stochastic process such that lim $N \to \infty$ $\overline{X}(N) = \mu$. Let $x(t)$ be a sample function (or realization) from this process. Then for any $\epsilon > 0$ and $\delta < 1$ exists an N_0 such that

(a) if $T = \{t; t = 0,1, \ldots\}$, then

$$\left| \frac{1}{N} \sum_{t=0}^{N-1} x(t) - \mu \right| < \epsilon, \qquad \text{when } N > N_0$$

with at least 100δ percent confidence, and

(b) if $T = \{t; 0 \leq t < \infty\}$, then

$$\left| \frac{1}{N} \int_0^N x(t)\, dt - \mu \right| < \epsilon, \qquad \text{when } N > N_0$$

with at least 100δ percent confidence.

Example 1. Let

$$X(t) = C + N(t), \qquad t \geq 0,$$

where C is a constant and $N(t)$ is random noise with the properties

$$E[N(t)] = 0 \tag{11}$$

and

$$R_N(t, t + \tau) = R_N(\tau) = C^2 \qquad \text{if } |\tau| > \tau_0. \tag{12}$$

Now suppose we observe $X(t)$ continuously and desire to estimate C. Clearly

$$E[X(t)] = E[C] + E[N(t)] = C.$$

Moreover, the process is stationary by equations (11) and (12) and equation (12) guarantees that it is ergodic in the mean by Corollary 1. Thus the sample average

$$\bar{X}(N) = \frac{1}{N} \int_0^N X(t)\,dt \longrightarrow C$$

as $N \longrightarrow \infty$.

We now turn our attention to the question of ergodicity of the autocorrelation. In regard to this we have the following results. We restrict ourselves to stochastic processes with index set $T = \{t;\ 0 \leq t < \infty\}$, although the results hold for the index set $T = \{t; 0,1, \ldots\}$ if integrals are replaced by sums.

Theorem 3 If for each fixed τ,

$$\psi_X(u) = E[X(t + \tau)X(t)X(t + \tau + u)X(t + \tau)]$$

is independent of t, then

$$\lim_{N \to \infty} \bar{R}_X(N; \tau) = R_X(\tau) \tag{13}$$

if and only if

$$\lim_{N \to \infty} \frac{1}{N} \int_0^N [\psi_X(u) - R_X^2(\tau)\ du = 0. \tag{14}$$

It is easily shown that (14) holds if

$$\lim_{\tau \to \infty} C_X(\tau) = 0$$

and thus (13) holds if the dependency between $X(t)$ and $X(t + \tau)$ wears off as τ gets large. It should be noted that this is the same sufficient condition we obtained for ergodicity in the mean.

Theorem 4 If $\{X(t); \; 0 \leq t < \infty\}$ and $\{Y(t); \; 0 \leq t < \infty\}$ are two jointly stationary processes, then

$$\lim_{N \to \infty} \bar{R}_{XY}(N; \tau) = R_{XY}(\tau) \tag{15}$$

if and only if

$$\lim_{N \to \infty} \frac{1}{N} \int_0^N [\psi_{XY}(u) - R_{XY}^2(\tau)] \, du = 0, \tag{16}$$

where

$$\psi_{XY}(u) = E[X(t + \tau)Y(t)X(t + \tau + u)Y(t + u)].$$

Equation (16) can be shown to hold if

$$\lim_{\tau \to \infty} C_{XY}(\tau) = 0$$

so that if $X(t + \tau)$ and $Y(t)$ are uncorrelated for large τ, (15) holds.

Before closing this section on estimation by time sample averages let us discuss the possibility of estimating the distribution function $F_{X(t)}(x)$, that is, the $P[X(t) \leq x]$, by a sample average. In regard to this we have the following:

Theorem 5 Let $\{X(t); \; 0 \leq t < \infty\}$ be a strictly stationary process and let

$$y(t) = \begin{cases} 1, & x(t) \leq x, \\ 0, & x(t) > x. \end{cases}$$

Then

$$\lim_{N\to\infty} \frac{1}{N} \int_0^N Y(t)\,dt = F(X) = P[X \leq x] \tag{17}$$

if and only if

$$\lim_{N\to\infty} \frac{1}{N} \int_0^N \left(1 - \frac{\tau}{T}\right)\{F[x,x;\tau] - F^2(x)\}\,d\tau = 0, \tag{18}$$

where

$$F[x,x;\tau] = P[X(t) \leq x, X(t+\tau) \leq x].$$

It is not difficult to show that (18) holds if

$$\lim_{\tau\to\infty} F[x,x;\tau] = F^2(x). \tag{19}$$

Note that if $X(t)$ and $X(t+\tau)$ are independent when τ is sufficiently large, then equation (19) holds and hence once more a sufficient condition is that the dependency between $X(t)$ and $X(t+\tau)$ wears off as τ gets large.

We point out again that the implication of theorems such as Theorem 5, in regard to observations, that is, a realization, is that the quantity of interest can be estimated to any degree of accuracy with any desired confidence by an appropriate selection of N. Thus in this case for any $\epsilon > 0$ and any desired confidence level, there is an N such that

$$\frac{1}{N} \int_0^N y(t)\,dt - F(x) < \epsilon$$

at the required level of confidence.

We should make a comment in regard to Theorem 5. The theorem is really quite intuitive. It simply says that for a strictly stationary process satisfying equation (18) we estimate for any t, $P[X(t) \leq x]$ by the proportion of time the graph of $x(t)$ remains below x. This is illustrated in Figure 6.5.

Finally let us comment in regard to the index sets, $T = \{t; t = 0,1,2,\ldots\}$ and $T = \{t; 0 \leq t < \infty\}$, which we have restricted ourselves to in this chapter. Similar results hold for

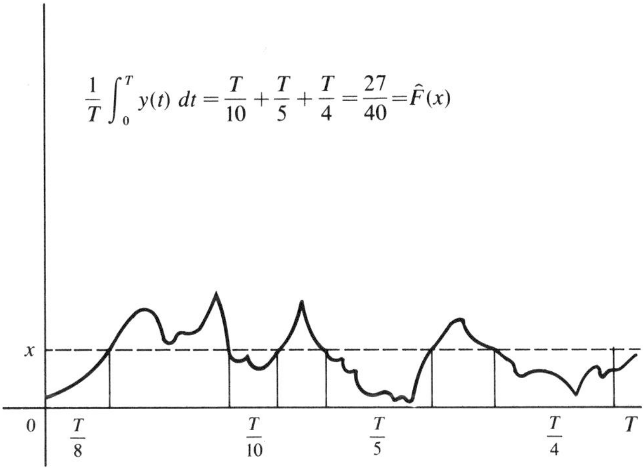

Figure 6.5

different index sets. For example, in Theorem 1, if we replace $T = \{t;\ 0 \leq t < \infty\}$ by $T = \{t;\ -\infty < t < \infty\}$, then the result is the same, but in this case,

$$\overline{X}(N) = \frac{1}{2N}\int_{-N}^{N} X(t)\,dt.$$

For similar results using $T = \{t;\ -\infty < t < \infty\}$, see Papoulis (Ref. 17).

Special Processes Most of the density functions discussed in Chapter IV have a corresponding interpretation as stochastic processes. We will now discuss a few of these briefly.

6.5 THE POISSON PROCESS

As we have seen in Section 4.6 the Poisson distribution arises as a probability model for describing the number of occurrences of a particular event, E, within a fixed time period when the following assumptions hold:

(a) $0 < P[E] < 1$.

(b) The events that E occurs in each of two nonoverlapping intervals are independent.

(c) The probability E occurs in a given interval is proportional to the interval length.

(d) The probability that E occurs more than once in the interval Δt approaches zero as $\Delta t \to 0$.

Now let $X(t)$ be a random variable whose values are defined by the number of times E occurs within the interval $(0, t)$. Then clearly $\{X(t); 0 \leq t < \infty\}$ is a stochastic process. Such a process is called an integer-valued process. If we assume that $X(t)$ has the appropriate Poisson distribution for each t, then it would seem quite natural to define it to be a Poisson process. This is essentially what is done. However, we should be more specific, so we now formally define a Poisson process. First it is necessary to give the following two definitions.

Definition 1 Let $\{X(t); 0 \leq t < \infty\}$ be a stochastic process. Then we say the process has independent increments if $X(t_1) - X(t_2), X(t_2) - X(t_3), \ldots, X(t_n) - X(t_{n-1})$ are independent random variables for every choice of $t_1, t_2, \ldots, t_n$ such that $t_1 < t_2 < t_3 < \cdots < t_n$.

Definition 2 If the process of Definition 1 has the additional property that $X(t_1 + h) - X(t_2 + h)$ has the same distribution for every $h \geq 0$, then we say that the process has stationary independent increments.

Note that Definition 1 is essentially assumption (b), while Definition 2 is of the same nature as assumption (c).

Definition 3 Let $\{X(t); 0 \leq t < \infty\}$ be an integer-valued process with stationary independent increments. The process is said to be a Poisson process if for every $s, t \in [0, \infty)$ such that $s \leq t$ the random variable $X(t) - X(s)$ has a Poisson distribution with mean $\mu(t) = (t - s)k$, k a constant; that is,

$$P[X(t) - X(s) = N] = \frac{e^{-(t-s)k}[(t - s)k]^N}{N!}. \tag{1}$$

From (2), Section 3.3, we have

$$E[X(t) - X(s)] = (t - s)k = \text{Var}[X(t) - X(s)].$$

We now state the following expected theorem. Its proof is effectively the same as the first motivation in Section 4.6.

Theorem 1 Let $\{X(t);\ 0 \leq t < \infty\}$ be a counting process with stationary independent increments for an event E satisfying assumptions (a) and (d) above. Then $\{X(t);\ 0 \leq t < \infty\}$ is a Poisson process.

Note that we have not included assumptions (b) and (c) directly. This is because they are included by the assumption of stationary independent increments. We have stated this theorem so that the reader can develop a feel for when a physical process might be considered a Poisson process.

Example 1. Consider again Example 1, Section 4.6. Clearly, if $X(t)$ is defined by the number of failures in the interval $(0, t]$ of the system of that example, then $\{X(t)\}$ is a Poisson process.

Example 2. In studying shot noise in a vacuum tube it becomes necessary to determine the probability $P[N; t]$ that exactly N electrons will be emitted by the cathode tube during a time interval of length t. If the temperature remains constant, it is reasonable to assume that the probability that an electron will be emitted in a given time interval is independent of the number previously emitted. Moreover, for short intervals it is also reasonable to assume that

$$P[N; t] = kt.$$

Consequently, $P[N; t]$ satisfies equation (1) and we have

$$P[X(t) - X(0) = N] = P[X(t) = N] = \frac{e^{-tk}(tk)^N}{N!}.$$

For more discussion of the Poisson process and some generalization of it, see Parzen (Ref. 19, Chap. 4).

6.6 THE DERIVATIVE PROCESS

Given a stationary stochastic process $\{X(t);\ 0 \leq t < \infty\}$, a new process $\{X'(t)\}$ can be defined by

$$X'(t) = \lim_{h \to 0} \frac{X(t + h) - X(t)}{h} \tag{1}$$

for all t such that the above limit exists in the sense that

$$\lim_{h\to 0} E\left[\left|\frac{X(t+h)-X(t)}{h} - X'(t)\right|^2\right] = 0. \tag{2}$$

The first- and second-order moments of the process $\{X'(t)\}$ can usually be obtained from the process $\{X(t)\}$ by means of the following theorem.

Theorem 1 Let $\{X(t);\ 0 \le t < \infty\}$ be a stationary process such that $R''_X(\tau)$ exists. Then

$$\mu_{X'(t)} = E[X'(t)] = \frac{d}{dt}E[X(t)] = 0, \tag{3}$$

$$\begin{aligned} C_{X'}(\tau) &= E\{[X'(t+\tau) - \mu'(t+\tau)][X'(t) - \mu'(t)]\} \\ &= -C''_X(\tau) \\ &= -R''_X(\tau) = R_{X'}(\tau), \end{aligned} \tag{4}$$

$$\begin{aligned} C_{X'X}(\tau) &= E\{[X'(t+\tau) - \mu'(t+\tau)][X(t) - \mu(t)]\} \\ &= R_{X'X}(\tau) \\ &= R'_X(\tau), \end{aligned} \tag{5}$$

$$\begin{aligned} C_{XX'}(\tau) &= R_{XX'}(\tau) \\ &= -R'_X(\tau). \end{aligned} \tag{6}$$

A stronger statement can be made in regard to (4). That is, it can be shown that $X'(t)$ exists if and only if $R''_X(\tau)$ exists. Thus it follows that if $X(t)$ is a stationary process, then $X'(t)$ is also a stationary process. The process $X'(t)$ has a number of interesting properties. We will discuss some of them in the following examples.

Example 1. Let $\{X(t);\ 0 \le t < \infty\}$ be a stationary process such that $X'(t)$ exists for each t. We will now show that $X(t)$ and $X'(t)$ are uncorrelated. The result follows from the fact that the existence of $X'(t)$ implies the existence $R''_X(\tau)$. But then by Theorem 1 we have

$$R'_X(0) = R_{X'X}(0) = R_{XX'}(0) = -R'_X(0),$$

or

$$R'_X(0) = 0.$$

Thus

$$E[X(t)X'(t)] = R_{XX'}(0) = 0,$$

as claimed.

Example 2. Suppose $E[X(t)] = \mu$ and $R_x(\tau) = e^{-\alpha\tau^2}$, where $\alpha > 0$. What can we infer in regard to $X'(t)$? The answer to this question is that we can determine $X'(t)$ through its second-order moments. Thus Example 1 is not to be construed as saying the process $\{X(t)\}$ conveys no information in regard to the process $\{X'(t)\}$. Note for this example we have

$$\begin{aligned} E[X'(t)] &= 0, \\ R_{X'}(\tau) &= \alpha e^{-\alpha\tau^2} - 4\alpha^2\tau^2 e^{-\alpha\tau^2}, \\ R_{XX'}(\tau) &= 2\alpha\tau e^{-\alpha\tau^2}, \\ R_{X'X}(\tau) &= -2\alpha\tau e^{-\alpha\tau^2}, \end{aligned}$$

and, of course,

$$R_{X'X}(0) = R_{XX'}(0) = 0.$$

Example 3. If $R_X(\tau) = e^{-\alpha|\tau|}$, what can one say about $X'(t)$?

Solution: $X'(t)$ does not exist, since $R''_X(\tau)$ does not exist at $\tau = 0$.

6.7 THE NORMAL PROCESS

A process $\{X(t); 0 \leq t < \infty\}$ such that for every finite subset the random variables $X(t_1), X(t_2), \ldots, X(t_n)$ have a multivariate normal distribution is called a normal process. One should recall that the multivariate normal distribution is completely determined by its mean and covariance. It follows, then, that a normal process is completely determined by $E[X(t)]$ and $\mathrm{Cov}[X(s), X(t)]$. Thus a weakly stationary normal process is also strictly stationary. The process plays a basic role in stochastic processes and is one of the most thoroughly studied.

The reason for its position of prominence is, of course, the central limit theorem. However, its mathematical tractability would have to be considered as playing a basic role in its development and at times its misuse.

One of the most important facets of a normal process from the engineer's point of view is the fact that if $\{X(t)\}$ is a normal process, then $\{X'(t)\}$, $\left\{\int_a^t X(\xi)\,d\xi\right\}$, and $\{X(t+h) - X(t)\}$ are also normal processes. These results can all be obtained essentially from the following fundamental theorem.

Theorem 1 Let$X(t_1)$, $X(t_2), \ldots, X(t_n)$ belong to a normal process and let

$$\begin{aligned} Y_1 &= \alpha_{11}X(t_1) + \alpha_{12}X(t_2) + \cdots + \alpha_{1n}X(t_n) \\ Y_2 &= \alpha_{21}X(t_1 + \alpha_{22}X(t_2) + \cdots + \alpha_{2n}X(t_n) \\ &\vdots \\ Y_m &= \alpha_{m1}X(t_1) + \alpha_{n2}X(t_2) + \cdots + \alpha_{mn}X(t_n). \end{aligned}$$

Then $Y_1, \ldots, Y_m$ are jointly normal with

$$E[Y_k] = \sum_{j=1}^{n} \alpha_{kj}E[X(t_j)]$$

and

$$\mathrm{Cov}[Y_j, Y_k] = \sum_{p,q=1}^{n} \alpha_{jp}\alpha_{qk}\mathrm{Cov}[x(t_p), x(t_q)].$$

Example 1 (Brownian Motion). The motion of small particles immersed in a gas or liquid is termed Brownian motion in honor of the botanist Robert Brown, who first discovered its existence. Since this motion is continuous in time and quite random in nature, it can be best interpreted as a stochastic process. Let us consider the nature of this process. First, let $X(t)$ denote the displacement of a particle from $X(0)$ in some three-dimensional coordinate system where we define $X(0) = 0$. Then $X(t) - X(s)$, $s < t$, is the displacement over the time interval $t - s$. Now it was shown by Einstein in 1905 that Brownian motion could be explained by considering the motion to be a result of a constant bombardment of molecules. Hence it is quite reasonable to assume that the total displacement of a particle during the time interval $s - t$, which is large in comparison to the time between impacts, is the sum of a large number of very small irregular displacements. Consequently, by the central limit theorem we can assume that $X(t) - X(s)$ is normal with zero mean. Moreover, from our discussion it is quite clear that the position of a particle at time t is independent of its position at time $t_0 < t$. Further, if we assume the system to be static, that is, in a state of equilibrium (constant temperature, pressure, etc.), then the distribution of $X(t) - X(s)$ should be the same as $X(t + h) - X(s + h)$. That is, it should certainly be independent of when we begin observation of the process. These latter comments are, of course, equivalent to the mathematical assumption of stationary independent increments. We have therefore established that if $\{X(t)\}$ is a Brownian process, it should have the following properties

(a) stationary independent increments.
(b) $X(t)$ is distributed normal for $t > 0$.
(c) $E[X(t)] = 0$.
(d) $X(0) = 0$.

Under assumptions (a)–(d) it is not difficult to show that a Brownian process is a normal process. Processes which satisfy conditions (a)–(d) are also often called Wiener processes, in honor of Norbert Wiener, who was one of the first to consider Brownian motion as a stochastic process satisfying conditions (a)–(d).

For further discussion of normal processes, see Papoulis (Ref. 17, Chap. 15).

VII
Linear Estimation of Stochastic Processes

7.1 INTRODUCTION

In Chapter VI we defined a stochastic process and established methods of estimating the first- and second-order moments of stationary processes. There is, however, an estimation problem of a different nature associated with stochastic processes. That is, given a stochastic process $\{X(t);\ t \in T\}$, determine a best estimate for a process $\{Y(t);\ t \in T\}$. Thus, for example, one may be observing $X(t)$ and desire to estimate $S(t)$, where

$$X(t) = S(t) + N(t). \tag{1}$$

Equation (1) differs from Example 1, Section 6.4, in that here $S(t)$ is a nonconstant random variable with constant expected value. Another example is that of observing $X(t)$ and desiring to estimate $X'(t)$.

The general problem of estimation of a stochastic process is a difficult one and one which has been studied extensively. The corpus of theory which has resulted is quite large and we will not attempt even to summarize it here. We will, however, give a brief introduction to the problem and direct the reader to more complete sources.

Throughout this chapter we will assume that all the processes involved are stationary and that their first- and second-order moments are known or at least have been satisfactorily estimated by such methods as those of the previous chapter or other appropriate means. Under these assumptions it is no loss in generality to further assume that all processes considered have zero means, since if μ_Y is known we can replace the problem of estimating $Y(t)$ by the problem of estimating $Y(t) - \mu_y$. Thus we will assume that all processes involved have zero means.

7.2 BEST ESTIMATORS

Given $X(\xi)$ for all $\xi \in I$, we desire to estimate $Y(t)$ for a fixed t in a manner which in some sense is desirable. Our measure of desirability will simply be expected squared error and the definition we now give for a best estimator is only slightly different than our more restricted definition of Chapter V.

Definition 1 For a fixed $t \in T$, $\hat{Y}(t) = H[X(\xi); \xi \in I]$ will be called a best estimator for $Y(t)$ if

$$E[|Y(t) - \hat{Y}(t)|^2] = \text{minimum}. \tag{1}$$

In this definition the notation $H[X(\xi); \xi \in I]$ is meant to imply that $\hat{Y}(t)$ is possibly a function of all the $X(\xi)$'s such that ξ is an element of some index set I. For example, if $X(\xi)$ were known for $\xi \in \{1,2,3\}$, then $\hat{Y}(t)$ would be a function of $X(1)$, $X(2)$, and $X(3)$, whereas if $X(\xi)$ were known for $\xi \in (1,3)$, then $\hat{Y}(t)$ might be a function of all $X(\xi)$ such that ξ is in that interval.

From a practical point of view the attempt to establish a best estimator is usually a little overambitious. We will discuss this further in the last section of this chapter. For the present, however, we avoid these difficulties by directing our attention to best linear estimators which we now define.

Definition 2 For a fixed $t \in T$, $\hat{Y}(t) = H[X(\xi); \xi \in I]$ will be called a best linear estimator for $Y(t)$ if

$$E[|Y(t) - \hat{Y}(t)|^2] = \text{minimum over } S, \tag{2}$$

where S is a set of all linear estimators of a specified form.

The fact that Definition 2 leads to some useful results is demonstrated by the next theorem.

Theorem 1 Let t be fixed and let $X(\xi)$ be known for $\xi = 1, \ldots, n$. Then a necessary and sufficient condition that

$$\hat{Y}(t) = \sum_{k=1}^{n} h_k X(k), \tag{3}$$

where the h_k's are constants, is a best linear estimator for $Y(t)$ of the form defined by (3) is that

$$E\left\{\left[Y(t) - \sum_{k=1}^{n} h_k X(k)\right] X(\xi)\right\} = 0 \qquad \text{for } \xi = 1,2,\ldots,n. \tag{4}$$

Moreover, the expected squared error, E.S.E., is given by

$$\begin{aligned} \text{E.S.E.} &= E\{[Y(t) - \hat{Y}(t)]^2\} \\ &= E\left\{\left[Y(t) - \sum_{k=1}^{n} h_k X(k)\right][Y(t)]\right\}. \end{aligned} \tag{5}$$

Note that since $Y(t)$ is possibly a different random variable for each t, that the coefficients h_k which satisfy equation (4) may be different for different values of t, so that the notation

$$\hat{Y}(t) = \sum_{k=1}^{n} h_k(t) X(k) \tag{6}$$

is probably better than that of equation (3).

Example 1 (The Problem of Prediction). We observe a stochastic process $\{X(t)\}$ at $t = t_1, t_2, \ldots, t_n$, where $t_1 < t_2 < \cdots < t_n$ and desire to estimate $X(t_n + \lambda)$, $\lambda > 0$, by means of an estimator of the form

$$\hat{X}(t_n + \lambda) = \sum_{k=1}^{n} h_k X(t_k). \tag{7}$$

We assume $E[X(t)] = 0$ for any t, as mentioned in the introduction. Determine the h_k's so that $\hat{X}(t_n + \lambda)$ is a best estimator for $X(t_n + \lambda)$.

Solution: By Theorem 1, we know that the h_k's satisfy the system of equations

$$E\left\{\left[X(t_n + \lambda) - \sum_{k=1}^{n} h_k X(t_k)\right] X(t_m)\right\} = 0, \qquad \text{for } m = 1,2,\ldots,n. \tag{8}$$

Now using the notation

$$E[X(s)X(r)] = R_X(s - r)$$

and noting that $R_X(s - r) = R_X(r - s)$, we can write the system of equations (8) as the system

$$\begin{aligned}
h_1 R_X(0) + h_2 R_X(t_2 - t_1) + \cdots + h_n R_X(t_n - t_1) &= R_X(t_n + \lambda - t_1) \\
h_1 R_X(t_2 - t_1) + h_2 R_X(0) + \cdots + h_n R_X(t_n - t_2) &= R_X(t_n + \lambda - t_2) \\
\vdots \qquad\qquad \vdots \qquad\qquad\qquad \vdots \qquad &\qquad\qquad \vdots \qquad\qquad (9) \\
h_1 R_X(t_n - t_1) + h_2 R_X(t_n - t_2) + \cdots + h_n R_X(0) &= R_X(\lambda).
\end{aligned}$$

Several points should be noted in regard to the above system of equations. First, note that if the $X(t)$'s are uncorrelated, then the system has only the trivial solution $h_1 = h_2 = \cdots = h_n = 0$. Thus equation (8) results in the very reasonable result that the best estimate for $X(t_n + \lambda)$ when it is uncorrelated with the previous elements of the process is $\hat{Y}(t) = E[X(t_n + \lambda)] = 0$. On the other hand, if $R_X(\lambda) \neq 0$, the system has a unique nontrivial solution. Further, if $n = 1$, the system becomes the simple equation

$$h_1 R_X(0) = R_X(\lambda) \tag{10}$$

or

$$h_1 = \frac{R_X(\lambda)}{R_X(0)}.$$

Now suppose that

$$R_X(\tau) = e^{-\alpha|\tau|}, \qquad \alpha > 0. \tag{11}$$

Then

$$h_1 = e^{-\alpha\lambda}$$

since $\lambda > 0$. Moreover,

$$\begin{aligned}
\text{E.S.E.} &= E\{[X(t_1 + \lambda) - h_1 X(t_1)][X(t_1 + \lambda)]\} \\
&= 1 - e^{-2\alpha\lambda}. \qquad (12)
\end{aligned}$$

Note that $R_X(0) = 1$, which is the variance of $X(t)$. However, for large α we see by (12) that E.S.E. ≈ 1, and hence, in this case, our estimator is of little value. Of course, when α is large, we see by (11) that we are essentially back to the uncorrelated case.

Finally note, for the autocorrelation defined by (11), that when $k = 1,2,\ldots,n$,

$$
\begin{aligned}
E\{[X(t_n + \lambda) - h_n X(t_n)]X(t_k)\} \\
&= R_X(t_n + \lambda - t_k) - h_n R_X(t_n - t_k) \\
&= e^{-\alpha(t_n - t_k + \lambda)} - h_n e^{-\alpha(t_n - t_k)} \\
&= 0
\end{aligned}
$$

if $h_n = e^{-\alpha\lambda}$. But in this case the system of equations (9) has the unique solution $h_n = e^{-\alpha\lambda}$ and $h_1 = h_2 = \cdots = h_{n-1} = 0$. Thus we see for a process with autocorrelation $ce^{-\alpha|\tau|}$ that if the present is known, the past is of no value in estimating the future. The converse of the above is also true. That is, if we have a stationary stochastic process in which the past is of no value in estimating the future when the present is known, then the autocorrelation of the process is of the form $ce^{-\alpha|\tau|}$. Such processes are called *wide-sense Markoff processes.*

Example 2 (The Filter Problem). Given three processes $\{X(t)\}$, $\{S(t)\}$, and $\{N(t)\}$, such that $E[X(t)] = E[S(t)] = E[N(t)] = 0$, and

$$X(t) = S(t) + N(t). \tag{13}$$

Find a best estimator for $S(t_0)$ of the form $hX(t_0)$, h a constant.

Solution: By Theorem 1, h must satisfy the equation

$$E\{[S(t_0) - hX(t_0)]X(t_0)\} = 0.$$

Thus letting $R_{SX}(\tau) = E[S(t + \tau)X(t)]$ we have

$$R_{SX}(0) = hR_X(0)$$

or

$$h = \frac{R_{SX}(0)}{R_X(0)}. \tag{14}$$

Example 3 (The Smoothing Problem). Suppose once more that

$$X(t) = S(t) + N(t).$$

Find a best estimator, $\hat{S}(t_2)$, for $S(t_2)$ of the form $h_1 X(t_1) + h_2 X(t_2) + h_3 X(t_3)$.

Solution: Again by Theorem 1 the h_i's must satisfy the system

$$\begin{aligned} h_1 R_X(0) + h_2 R_X(t_2 - t_1) + h_3 R_X(t_3 - t_1) &= R_{SX}(t_2 - t_1), \\ h_1 R_X(t_2 - t_1) + h_2 R_X(0) + h_3 R_X(t_3 - t_2) &= R_{SX}(0), \\ h_1 R_X(t_3 - t_1) + h_2 R_X(t_3 - t_2) + h_3 R_X(0) &= R_{SX}(t_3 - t_2). \end{aligned}$$

Solving the above system gives the desired estimator. Note that if the $X(t_i)$ are uncorrelated, the estimator becomes the one of Example 2. In that case the above system has the solution $h_1 = h_3 = 0$ and

$$h_2 = \frac{R_{SX}(0)}{R_X(0)},$$

so that

$$\hat{S}(t_2) = \frac{R_{SX}(0)}{R_X(0)} X(t_2).$$

For a more complete discussion of these and other examples see Papoulis (Ref. 17, pp. 394–400).

Now note that the constants $h_i(t)$ which determine the optimum estimator are functions of the ξ_i, that is, the points where the $X(t)$'s are known. Then we see that the estimator

$$\hat{Y}(t) = \sum_{k=1}^{n} h(\xi_k; t) X(\xi_k)\, \Delta\xi_k, \tag{15}$$

where $\Delta\xi_k = \xi_{k+1} - \xi_k$, is an estimator of the form given in equation (6), where the constants $h_k(t) = h(\xi_k; t)\, \Delta\xi_k$.

Suppose, however, that $X(\xi)$ is known for all ξ in some interval (a, b). An estimator such as given by (15) clearly does not make use of all the available information in this case. It therefore seems natural to adopt a new form for the estimator

$\hat{Y}(t)$. Let

$$\hat{Y}(t) = \lim_{n \to \infty} \sum_{k=1}^{n} h(\xi_k; t) X(\xi_k) \Delta \xi_k,$$

$$\text{where } a = \xi_1 < \xi_2 < \cdots < \xi_n = b$$

or

$$\hat{Y}(t) = \int_a^b h(\xi; t) X(\xi)\, d\xi. \tag{16}$$

Any physical or mathematical system in which $\hat{Y}(t)$ is described by (16) is called a *linear system*. The function $\hat{Y}(t)$ is called the *output*, $X(t)$ is called the *input*, and $h(\xi; t)$ is called the *weighting function* or the *impulse response function*. From the above it is clear that the determination of the impulse response function is directly analogous to the problem solved by Theorem 1. That is, it is the continuous-data analogue to selecting the coefficients in a linear combination of the $X(\xi)$'s so as to minimize the expected squared error. In regard to this we have the following theorem.

Theorem 2 Let t be fixed and let $X(\xi)$ be known for $\xi \in (a, b)$. Then a necessary and sufficient condition that

$$\hat{Y}(t) = \int_a^b h(u; t) X(u)\, du \tag{17}$$

is a best estimator for $Y(t)$ of the form defined by (17) is

$$E\left\{\left[Y(t) - \int_a^b h(u; t) X(u)\, du\right] X(\xi)\right\} = 0, \tag{18}$$

for every $\xi \in (a, b)$.

Moreover, when (18) holds,

$$\begin{aligned} \text{E.S.E.} &= E\{[Y(t) - \hat{Y}(t)]^2\} \\ &= E\left\{\left[Y(t) - \int_a^b h(u; t) X(u)\, du\right] Y(t)\right\}. \end{aligned} \tag{19}$$

Figure 7.1 gives a "black-box" interpretation of Theorem 2. Note that the system is completely determined by $h(\xi; t)$, which follows from (18). In terms of correlation functions we can

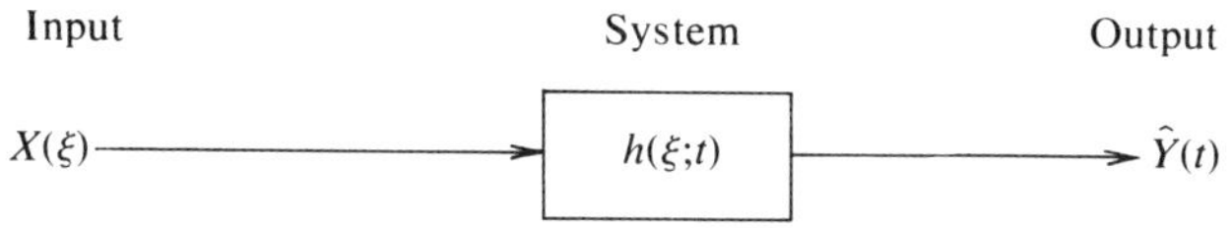

Figure 7.1 Linear System

write (18) and (19) in the forms

$$R_{YX}(t - \xi) = \int_a^b h(u;t) R_X(u - \xi)\, du \tag{20}$$

and

$$\text{E.S.E.} = R_Y(0) - \int_a^b h(u;t) R_{XY}(u - t)\, du, \tag{21}$$

respectively, so that the problem of determining $h(u;t)$ is one of solving the integral equation (20) for all $\xi \in (a, b)$. The estimator defined by (17) is often called a linear filter. This is also true for the estimator of Theorem 1. Example 2 discusses the so-called filter problem. It is in connection with this problem that the estimator acts properly as a filter, that is, filters signal from noise. However, it is often referred to as a filter in other types of problems and we will sometimes refer to it as such. Note that the integral in (21) serves as a measure of the effectiveness of the filter. That is, if the filter (or estimator) is to be of value, we should have E.S.E. $< R_Y(0)$, since $R_Y(0) = E\{[Y(t)]^2\}$, which is the expected error when simply estimating $Y(t)$ by zero, the mean of the process. From (21) it is therefore clear (and is intuitively obvious) that if $X(t)$ and $Y(t)$ are uncorrelated, the linear filter is of no value in estimating $Y(t)$.

7.3 LINEAR SYSTEMS WITH NONCAUSAL IMPULSE RESPONSE FUNCTIONS

Consider the linear system defined by

$$\hat{Y}(t) = \int_{-\infty}^{\infty} h(u;t) X(u)\, du. \tag{1}$$

In this case it can easily be shown that if both $\{\hat{Y}(t)\}$ and $\{X(t)\}$ are stationary processes, then $h(u; t)$ is a function of $t - u$ alone. Hence, since we are assuming all processes are stationary, we can write for (1),

$$\hat{Y}(t) = \int_{-\infty}^{\infty} h(t - u) X(u)\, du$$

$$= \int_{-\infty}^{\infty} X(t - u) h(u)\, du. \tag{2}$$

The integral in (2) is called the *convolution integral* and is denoted by $X * h(t)$. Thus we have

$$\hat{Y}(t) = X * h(t) \tag{3}$$

and

$$\hat{Y}(t) = h * X(t). \tag{4}$$

Now from (20), Section 7.2, we see that $\hat{Y}(t)$ is a best linear estimator if h satisfies the equation

$$R_{YX}(t - \xi) = \int_{-\infty}^{\infty} h(t - u) R_X(u - \xi)\, du$$

$$= \int_{-\infty}^{\infty} h(u) R_X(t - u - \xi)\, du$$

or

$$R_{YX}(\tau) = \int_{-\infty}^{\infty} h(u) R_X(\tau - u)\, du$$

$$= h * R_X(\tau). \tag{5}$$

In the above we have put no restriction on h other than assuming that the integrals involved exist as improper Riemann integrals. In many cases we will not even insist on this, but require the weaker assumption that the integrals involved can be given meaning as integrals of generalized functions. This should introduce no difficulties, since most engineers are well acquainted with such facts as

$$\int_{-\infty}^{\infty} R(\tau - u)\delta(u)\,du = R(\tau), \tag{6}$$

where $\delta(u)$ is the Dirac delta function, even though the statement is obviously false, if we do not allow a broader interpretation than the improper Riemann integral. For the justification of such results as (6) see Lighthill (Ref. 14) or Papoulis (Ref. 17).

Since we have put no restriction on $h(t)$ we must allow $h(t)$ to be nonzero if $t < 0$, and thus we say $h(t)$ is a noncausal function. That is, a function h is said to be a causal function if and only if $h(t) = 0$ for all $t < 0$.

The solution of equation (5) for $h(t)$ is quite easy to obtain by making use of the Fourier transform. [Comment: The reader who is unfamiliar with the Fourier transform should familiarize himself with the rudiments of that theory, since it is a basic tool in the analysis of linear systems. See Lighthill (Ref. 14) or Papoulis (Ref. 17).]

Now denote the Fourier transforms of $R_{XY}(\tau)$, $h(\tau)$, and $R_X(\tau)$ by $S_{YX}(\omega)$, $H(\omega)$, and $S_X(\omega)$, respectively. Then by taking the Fourier transform of both sides of (5), we have for the optimum filter

$$S_{YX}(\omega) = H(\omega)S_X(\omega). \tag{7}$$

Consequently, if $S_X(\omega) \neq 0$, we have

$$H(\omega) = \frac{S_{YX}(\omega)}{S_X(\omega)}. \tag{8}$$

Thus, by the Fourier inversion integral we have, if

$$h(t) = \frac{1}{2\pi}\int_{-\infty}^{\infty} e^{-i\omega t}H(\omega)\,d\omega, \tag{9}$$

$\hat{Y}(t)$ is a best linear estimator for $Y(t)$. Although the filter defined by (9) is not physically realizable, it is of some practical value and can certainly be approximated by realizable filters. For further discussion see Papoulis (Ref. 17, Chap. 11).

Example 1. Consider again the filtering problem. That is, let $Y(t) = S(t)$ and $X(t) = S(t) + N(t)$ and find a best estimator for $S(t)$. In this example, we assume $X(t)$ is known for $-\infty < t < \infty$ and we desire a best estimator $\hat{S}(t)$ of the form

$$\hat{S}(t) = \int_{-\infty}^{\infty} X(t-u)h(u)\,du. \tag{10}$$

Solution: By (8), $\hat{S}(t)$ is a best estimator for $S(t)$ if

$$H(\omega) = \frac{S_{SX}(\omega)}{S_X(\omega)},$$

and the expected squared error is given by

$$\text{E.S.E.} = R_S(0) - \int_{-\infty}^{\infty} R_{SX}(u)h(u)\,du. \tag{11}$$

An alternative form for (11) which is easily shown by means of the theory of Fourier transforms is

$$\text{E.S.E.} = \frac{1}{2\pi}\int_{-\infty}^{\infty}\left[S_S(\omega) - \frac{S_{SX}(\omega)S_{SX}(-\omega)}{S_X(\omega)}\right]d\omega. \tag{12}$$

Now suppose $S(t)$ and $N(t)$ are uncorrelated. Then $R_{SN}(\tau) = 0$, and therefore, $S_{SN}(\omega) = 0$. Also

$$S_{SX}(\omega) = \int_{-\infty}^{\infty} e^{i\omega\tau}R_{SX}(\tau)\,d\tau \tag{13}$$

and

$$S_X(\omega) = \int_{-\infty}^{\infty} e^{i\omega\tau}R_X(\tau)\,d\tau. \tag{14}$$

But

$$\begin{aligned} R_{SX}(\tau) &= E[S(t+\tau)X(t)] \\ &= E[S(t+\tau)S(t) + S(t+\tau)N(t)] \\ &= E[S(t+\tau)S(t)] \\ &= R_S(\tau) \end{aligned} \tag{15}$$

and

$$\begin{aligned} R_X(\tau) &= E[X(t+\tau)X(t)] \\ &= E[S(t+\tau)+N(t+\tau)][S(t)+N(t)] \\ &= R_S(\tau)+R_N(\tau). \end{aligned} \tag{16}$$

Combining (13), (14), (15), and (16) we have

$$S_{SX}(\omega) = S_S(\omega)$$

and

$$S_X(\omega) = S_S(\omega) + S_N(\omega).$$

Thus

$$H(\omega) = \frac{S_S(\omega)}{S_S(\omega)+S_N(\omega)}$$

and from (12) and the fact that for real processes (which is all we have considered) $S(\omega) = S(-\omega)$ we have

$$\text{E.S.E.} = \frac{1}{2\pi}\int_{-\infty}^{\infty} \frac{S_S(\omega)S_N(\omega)}{S_S(\omega)+S_N(\omega)}\,d\omega.$$

Thus if

$$S_S(\omega)S_N(\omega) = 0, \tag{17}$$

but

$$S_S(\omega)+S_N(\omega) \neq 0, \tag{18}$$

then

$$\text{E.S.E.} = 0. \tag{19}$$

Equations (17) and (18) are satisfied if the nonzero ranges of $S_S(\omega)$ and $S_N(\omega)$ are distinct, as indicated in Figure 7.2. When this is the case we have

$$H(\omega) = \begin{cases} 1 & \text{if } S_N(\omega) = 0, \\ 0 & \text{if } S_S(\omega) = 0, \end{cases} \tag{20}$$

and

$$\text{E.S.E.} = 0.$$

Consequently, the filter defined by (20) is a perfect filter; that is, it passes the signal and filters the noise, so that for the input

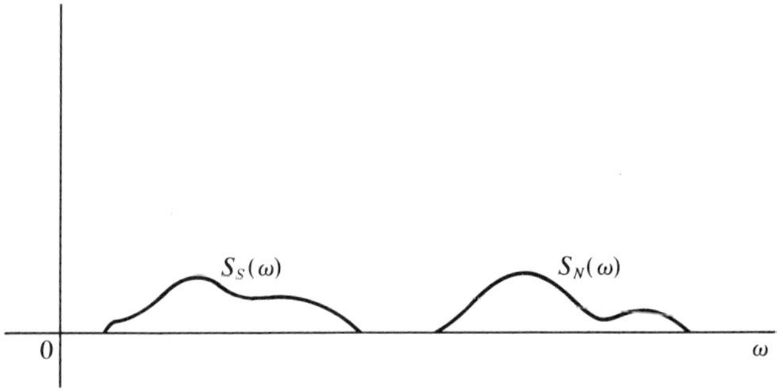

Figure 7.2 Spectral Densities

$X(t)$ the filter output is $S(t)$. A filter such as defined by (20) is called a bandpass filter.

7.4 LINEAR SYSTEMS WITH CAUSAL IMPULSE RESPONSE FUNCTIONS

Suppose now we assume that $X(\xi)$ is known for all $\xi \in (-\infty, t]$. Then we would desire a linear estimator of the form

$$\hat{Y}(t) = \int_{-\infty}^{t} X(u)h(t - u)\, du, \tag{1}$$

$$= \int_{0}^{\infty} X(t - u)h(u)\, du. \tag{2}$$

But this is the same as (2), Section 7.3, with the constraint that $h(t) = 0$ for $t < 0$. Thus (2) defines a linear system with a causal impulse response function. However, it is not necessary to look at the problem as an estimation problem subject constraints since we can interpret (2) in the sense of (1), and hence $\hat{Y}(t)$ is a best linear estimator if h satisfies the equation

$$R_{YX}(t - \xi) = \int_{-\infty}^{t} R_X(u - \xi)h(t - u)\, du, \qquad \xi \leq t,$$

that is,

$$R_{YX}(\tau) = \int_0^\infty R_X(\tau - u)h(u)\,du, \qquad \tau \geq 0. \tag{3}$$

Note that solving (3) for $h(t)$ is a much more difficult task than solving (5), Section 7.3. This is true since (3) only holds for $\tau \geq 0$, and hence we cannot take the Fourier transform of both sides of (3) as we did in (5), Section 7.3.

Example 1. Let $Y(t) = X(t + \alpha)$, $\alpha > 0$. Then the problem of estimating $Y(t)$ becomes one predicting $X(t + \alpha)$ if $X(\xi)$ is known for $\xi \leq t$. Thus

$$\hat{X}(t + \alpha) = \int_0^\infty X(t - u)h(u)\,du$$

is a best estimator (or optimum filter) for $X(t + \alpha)$ if $h(u)$ satisfies the integral equation

$$R_X(\tau + \alpha) = \int_0^\infty R_X(\tau - u)h(u)\,du. \tag{4}$$

Equation (4) is called the *Weiner–Hopf integral equation* and has been the subject of much investigation. For a discussion of its solution see Davenport and Root (Ref. 7) or Papoulis (Ref. 17).

In Example 1, Section 7.3, we saw that a perfect filter could be constructed under certain conditions when the signal and noise were uncorrelated. This example actually showed us much more—that the evaluation and construction of filters might more easily be accomplished through analysis of the functions $H(\omega)$, $S_X(\omega)$, and $S_{YX}(\omega)$, which are referred to as the system response function, the spectral density, and the cross-spectral density, respectively. This is, in fact, the case, and in addition these functions play a somewhat broader role in the analysis of stochastic processes. For this reason we devote the next section to their discussion.

7.5 THE SPECTRAL DENSITY AND SYSTEM RESPONSE FUNCTION

Definition 1 Let $\{X(t)\}$ be a stationary stochastic process with autocorrelation function $R_X(\tau)$ (see Figure 7.3).

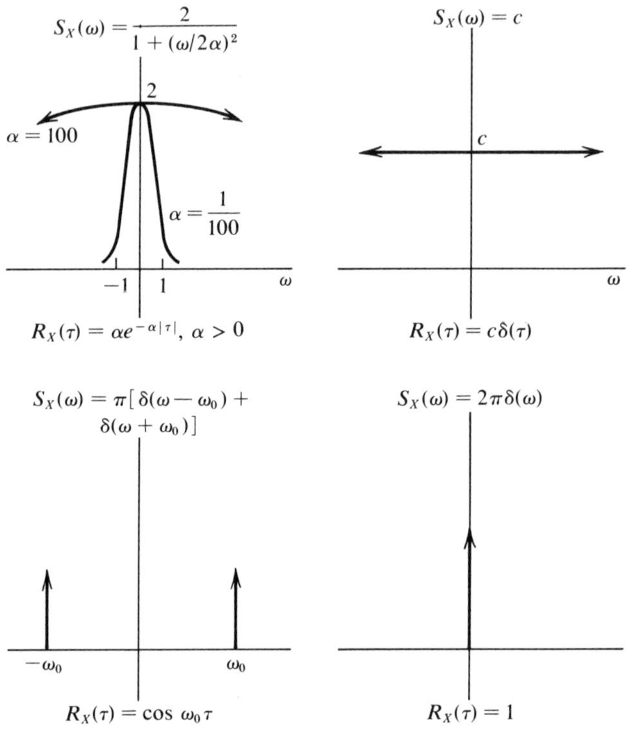

Figure 7.3 Spectral Density Plots for the Given Autocorrelation Function

Then we define the *spectral density* S_X of the process as the Fourier transform of $R_X(\tau)$, that is,

$$S_X(\omega) = \int_{-\infty}^{\infty} e^{i\omega\tau} R_X(\tau)\, d\tau. \tag{1}$$

Note that by the Fourier inversion formula

$$R_X(\tau) = \frac{1}{2\pi} \int_{-\infty}^{\infty} e^{-i\tau\omega} S_X(\omega)\, d\omega. \tag{2}$$

Because of relations (1) and (2) $S_X(\omega)$ and $R_X(\tau)$ are often called *Fourier transform pairs.* If $R_X(\tau)$ is considered as a generalized function on the appropriate test space, equations (1) and (2) hold without exception. We will assume that this property always holds [see Lighthill (Ref. 14)].

Note from (2) that

$$\sigma_X^2 = R_X(0) = \frac{1}{2\pi} \int_{-\infty}^{\infty} S_X(\omega)\, d\omega. \tag{3}$$

Thus we see that the total area under $S_X(\omega)$ is equal to 2π times the variance of the process and is therefore nonnegative. It can be shown that the stronger statement $S_X(\omega) \geq 0$ for all ω is also true.

Since $R_X(\tau)$ is an even function, that is, $R_X(\tau) = R_X(-\tau)$, equations (1) and (2) can be written in the alternative forms

$$S_X(\omega) = \int_{-\infty}^{\infty} R_X(\tau) \cos \omega\tau \, d\tau \tag{4}$$

and

$$R_X(\tau) = \frac{1}{2\pi} \int_{-\infty}^{\infty} S_X(\omega) \cos \omega\tau \, d\omega. \tag{5}$$

From equation (4) we see that $S_X(\omega) = S_X(-\omega)$ and also that $S_X(\omega)$ is a real function of ω.

Example 1. Let $\{\alpha(k);\ k = 1,\ 2, \ldots, n\}$ and $\{\beta(k);\ k = 1, 2, \ldots, n\}$ be stochastic processes such that the random variables $\alpha(1), \ldots, \alpha(n), \beta(1), \ldots, \beta(n)$ are uncorrelated, have zero means, and $E[\alpha^2(k)] = E[\beta^2(k)] = \sigma_k^2$ for each k.

Now let

$$X(t) = \sum_{k=1}^{n} \{\alpha(k) \cos \omega_k t + \beta(k) \sin \omega_k t\}. \tag{6}$$

Then clearly $E[X(t)] = 0$ and

$$R_X(\tau) = E[X(t + \tau) X(t)]$$

$$= E\left\{ \left[\sum_{k=1}^{n} \alpha(k) \cos \omega_k (t + \tau) + \beta(k) \sin \omega_k (t + \tau) \right] \right.$$

$$\times\left[\sum_{k=1}^{n} \alpha(k)\cos\omega_k t + \beta(k)\sin\omega_k t\right]\Bigg\}$$

$$= \sum_{k=1}^{n} \{E[\alpha^2(k)]\cos\omega_k(t+\tau)\cos\omega_k t$$

$$+ E[\beta^2(k)]\sin\omega_k(t+\tau)\sin\omega_k t\}$$

$$= \sum_{k=1}^{n} \sigma_k^2 \cos\omega_k\tau, \tag{7}$$

so that $\{X(t)\}$ is a stationary process. Moreover,

$$S_X(\omega) = \int_{-\infty}^{\infty} e^{i\omega\tau} R_X(\tau)\,d\tau = \pi\sum_{k=1}^{n} \sigma_k^2[\delta(\omega - \omega_k) + \delta(\omega + \omega_k)]. \tag{8}$$

Equation (6) has an interesting interpretation as a random Fourier series. The analogy between (6) and the Fourier expansion of a deterministic function is quite obvious. Here we have simply replaced the Fourier coefficients by random variables. Note that if we are willing to assume that the random variables $X(t)$ can be represented in such a manner, then $S_X(\omega)$ relates to us the harmonic content of $X(t)$. That is, it tells us precisely the nonzero frequencies in the representation of $X(t)$. Thus one might expect if $X(t)$ is approximately of the form (6) that a measurement (or calculation) of the spectral density should yield good approximations to the frequency composition of $X(t)$. But as we saw in Example 1, Section 7.3, this may be enough information to construct a satisfactory filter in a given situation. The analysis of the frequency composition of data is, in fact, the principal role of the spectral density in data processing.

It should be noted that we have established methods in Chapter VI for estimating $R_X(\tau)$, a quantity from which we can calculate $S_X(\omega)$. Of course, the last statement covers a lot of ground in which, needless to say, there are quite a few "stones." The reader is referred to Blackman and Tukey (Ref. 5).

Definition 2 Let $h(t)$ be the impulse response function for the linear system defined by (2), Section 7.3. We define the system response function $H(\omega)$ as the Fourier transform of $h(t)$; that is,

$$H(\omega) = \int_{-\infty}^{\infty} e^{i\omega t} h(t)\, dt. \tag{9}$$

As we have already seen, the system response function more or less serves as a link relating the observed process to the estimator process. This is, of course, quite reasonable, since $H(\omega)$ uniquely determines the weighting function $h(t)$ and vice versa. Now the weighting function is only a function of $R_{XY}(\tau)$ and $R_X(\tau)$, as can be seen by (7), Section 7.3, or (20), Section 7.2. Thus a number of important relations evolve between these functions and $H(\omega)$. The most important of these is given by the following theorem.

Theorem 1 Let $\hat{Y}(t)$ be the output of the linear system defined by (2), Section 7.3, with input $X(t)$ and system function $H(\omega)$. Then

$$S_{\hat{Y}}(\omega) = S_X(\omega) \,|\, H(\omega) \,|^2. \tag{10}$$

The proof of equation (10) is quite simple and follows immediately from the relations

$$S_{X\hat{Y}}(\omega) = S_X(\omega) H^*(\omega) \qquad \text{and} \qquad S_{\hat{Y}}(\omega) = S_{X\hat{Y}}(\omega) H(\omega),$$

where $H^*(\omega)$ is the complex conjugate of $H(\omega)$.

Example 2. Let

$$X(t) = S(t) + N(t)$$

as in Example 1, Section 7.3, and again suppose we desire to determine the system function [and hence the weight $h(t)$] so as to filter the noise $N(t)$ and pass the signal $S(t)$. Then from that example, we know

$$S_{SX}(\omega) = S_S(\omega)$$

and

$$S_X(\omega) = S_S(\omega) + S_N(\omega).$$

By (10),

$$S_{\hat{S}}(\omega) = S_S(\omega) \,|\, H(\omega) \,|^2 + S_N(\omega) \,|\, H(\omega) \,|^2. \tag{11}$$

Now we should note that even if $S_S(\omega)S_N(\omega) \neq 0$, it still may be possible to filter some of the noise. That is, if we can have $H(\omega)$ small at the dominant frequencies of the noise and close to unity at the dominant frequencies of $S(t)$, then the filter will pass most of the signal and filter most of the noise. Most stochastic processes have a Fourier-type representation similar to (6) [see Papoulis, Ref. 17, Chap. 14]. For this reason, in attempting to estimate or analyze a stochastic process it makes sense to talk about its frequencies. In fact, frequency characteristics turn out to be the best way to describe certain processes. We shall discuss some of these now.

7.6 WHITE NOISE

Consider the graphs of $R_X(\tau) = \alpha e^{-\alpha|\tau|}$ shown in Figure 7.4. Clearly for large α, $X(t)$ and $X(t + \tau)$ are essentially uncorrelated if τ is much greater than zero. If we consider the

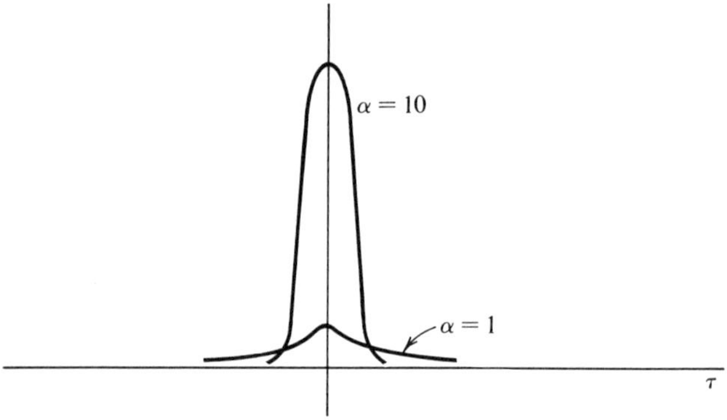

Figure 7.4 Autocorrelation Function

limit of $R_X(\tau)$ as $\alpha \longrightarrow \infty$, we could in some cases consider $R_X(\tau)$ as an appropriate autocorrelation for a process $\{X(t); 0 \leq t < \infty\}$ such that each of the $X(t)$ are uncorrelated or even independent if the process is normal. But

$$\lim_{\alpha \to \infty} R_X(\tau) = 2\delta(\tau).$$

Now let $\{X(t); 0 \leq t \leq \infty\}$ be a stochastic process with autocorrelation $R_X(\tau)$ defined by

$$R_X(\tau) = c\delta(\tau), \tag{1}$$

where c is a positive constant. Then

$$S_X(\omega) = c, \qquad -\infty < \omega < \infty. \tag{2}$$

Definition 1 When $S_X(\omega) = c, \quad -\infty < \omega < \infty,$ we call the stationary process $\{X(t)\}$ white noise.

From the motivation preceding this definition it is clear that white noise is for a continuous index set, an analogue of uncorrelated noise for a discrete index set.

Although white noise is in reality a hypothetical quantity, it serves as a good mathematical model for a number of physical processes, thermal noise and shot noise, being good examples.

Example 1. Let the input to a causal linear system with unknown weighting function be $X(t)$ and suppose $X(t)$ is white noise. Then if $\hat{Y}(t)$ is the output of the system, we have

$$\hat{Y}(t) = \int_0^\infty X(t - u)h(u)\,du$$

and multiplying both sides by $X(t + \tau)$ we obtain

$$X(t + \tau)\hat{Y}(t) = \int_0^\infty X(t + \tau)X(t - u)h(u)\,du.$$

Thus

$$\begin{aligned} R_{X\hat{Y}}(\tau) &= \int_0^\infty R_X(\tau + u)h(u)\,du \\ &= \int_0^\infty c\delta(\tau + u)h(u)\,du \\ &= ch(-\tau), \; -\infty < \tau < \infty, \end{aligned} \tag{3}$$

and therefore

$$R_{X\hat{Y}}(\tau) = 0 \qquad \text{if } \tau > 0;$$

that is, the input and output are uncorrelated for $\tau > 0$ and any weighting function. Now note that by (3)

$$ch(\tau) = R_{X\hat{Y}}(-\tau).$$

Thus it is possible to estimate $h(\tau)$ by estimating $R_{X\hat{Y}}(\tau)$ for negative τ. But when Theorem 4 of Chapter VI holds, this can be done by a time average; that is,

$$\bar{R}_{X\hat{Y}}(\tau) = \frac{1}{T}\int_0^T X(t+\tau)\hat{Y}(t)\,dt. \tag{4}$$

Equation (4) therefore defines a consistent estimator $\hat{h}(t)$ for $h(t)$, $t > 0$, by time averaging. That is,

$$c\hat{h}(t) = \frac{1}{T}\int_0^T X(\tau - t)\hat{Y}(\tau)\,d\tau.$$

7.7 BAND-LIMITED NOISE

Definition 1 Let $\{X(t)\}$ be a stationary stochastic process such that

$$S_X(\omega) = 0, \qquad \text{if } \omega \notin (\omega_0, \omega_1).$$

Then $\{X(t)\}$ is called a band-limited process.

One of the most important properties of band-limited processes is the following theorem.

Theorem 1 Let $\{X(t)\}$ be a stochastic process such that

$$S_X(\omega) = 0 \qquad \text{if } \omega > |\omega_0|. \tag{1}$$

Then

$$X(t) = \sum_{n=-\infty}^{\infty} X(nT)\frac{\sin(\omega_0 t - n\pi)}{\omega_0 t - n\pi}, \tag{2}$$

where

$$T = \pi/\omega_0.$$

The above theorem is called the *sample theorem* and the band-pass process defined by (1) is called a *low-pass process*. The theorem is the random-variable analogue of Shannon's theorem for deterministic signals. It tells us that if a stochastic process with continuous index set is a low-pass process, then it can be exactly reproduced from only a countable number of the random variables which make up the uncountable process.

Example 1. White noise can be thought of as a limiting case of band-limited noise, as follows. Let

$$S_X(\omega) = \begin{cases} c, & |\omega| < \omega_0, \\ 0, & \text{elsewhere}. \end{cases} \tag{3}$$

Then

$$\begin{aligned} R_X(T) &= \frac{c}{2\pi} \int_{-\omega_0}^{\omega_0} e^{-i\omega\tau}\, d\omega \\ &= \frac{c}{\pi} \int_0^{\omega_0} \cos \omega\tau \, d\omega \\ &= \frac{c\omega_0}{\pi} \left(\frac{\sin \omega_0 \tau}{\omega_0 \tau} \right). \end{aligned} \tag{4}$$

Since (4) approaches $c\delta(\tau)$ as $\omega_0 \longrightarrow \infty$, we can see by (3) or (4) that such a process is an approximation of white noise. Processes of this type are called band-limited white-noise processes.

7.8 CONCLUDING REMARKS

Given a portion of a stocha :ic process $\{X(\xi);\ \xi \in I \subset T\}$, we defined a best estimator for a stochastic process $\{Y(t)\}$ as one which minimized $E[\,|Y(t) - \hat{Y}(t)|^2]$. We did not, however, pursue this type of estimator but instead restricted ourselves to best linear estimators. It is possible to find the best estimator for $Y(t)$, theoretically. In fact, it can be shown that $\hat{Y}(t)$ is a best estimator for $Y(t)$ if

$$\hat{Y}(t) = E\{Y(t) \mid X(\xi), \xi \in I\}. \tag{1}$$

The difficulty in making any practical use of (1) is clear. That is, to obtain $\hat{Y}(t)$ from (1) we must know the joint density of $Y(t)$ and $X(\xi)$ for all $\xi \in I$. The latter is not generally possible. It can, however, be shown that if $Y(t)$ and $X(\xi)$ are multivariate normal for all $\xi \in I$, then $Y(t)$ as defined by (1) is a linear function of the $X(\xi)$ and hence is not only the best linear estimator but also the best estimator.

References and Selected Readings

1. Aitchison, John, and A. C. Brown, *Lognormal Distributions*, Cambridge University Press, New York, 1957.
2. Anderson, T. W., *An Introduction to Multivariate Statistical Analysis*, John Wiley & Sons, Inc., New York, 1958.
3. Bendat, J. S., *Principles and Applications of Random Noise Theory,* John Wiley & Sons, Inc., New York, 1958.
4. Bendat, J. S., and A. G. Piersol, *Measurement and Analysis of Random Data*, John Wiley & Sons, Inc., New York, 1966.
5. Blackman, R. B., and J. W. Tukey, *The Measurement of Power Spectra*, Dover Publications, Inc., New York, 1958.
6. Brunk, H. D., *An Introduction to Mathematical Statistics*, Blaisdell Publishing Co., Waltham, Mass., 2nd ed., 1965.
7. Davenport, W. B., and W. L. Root, *Random Signals and Noise*, McGraw-Hill, Inc., New York, 1958.
8. Erdélyi, A., *Higher Transcendental Functions*, Vol. 1, McGraw-Hill, Inc., New York, 1953.
9. Feller, W., *An Introduction to Probability Theory and Its Applications*, Vol. I, John Wiley & Sons, Inc., New York, 2nd ed., 1957.
10. Graybill, F. A., *An Introduction to Linear Statistical Models*, Vol. I, McGraw-Hill, Inc., New York, 1961.
11. Harter, H. L., *New Tables of the Incomplete Gamma-Function Ratio and of Percentage Points of the Chi-Square and Beta Distributions*, Aerospace Research Laboratories, U.S. Air Force, 1964 (available from Superintendent of Documents, U.S. Government Printing Office, Washington, D.C.).
12. Kendall, M. G., and A. Stuart, *The Advanced Theory of Statistics*, Vol. 1, Hafner Publishing Co., New York, 1958.
13. Lee, C. K., *Optimal Estimation, Identification, and Control,* M.I.T. Press, Cambridge, Mass., 1964.

14. Lighthill, M. J., *Introduction to Fourier Analysis and Generalized Functions*, Cambridge University Press, New York, 1964.
15. Meyer, P. L., *Introductory Probability and Statistical Applications*, Addison-Wesley Publishing Company, Reading, Mass., 1965.
16. Mood, A. M., and F. A. Graybill, *Introduction to the Theory of Statistics*, McGraw-Hill, Inc., New York, 2nd ed., 1963.
17. Papoulis, A., *Probability, Random Variables and Stochastic Processes*, McGraw-Hill, Inc., New York, 1965.
18. Parzen, Emanuel, *Modern Probability Theory*, John Wiley & Sons, Inc., New York, 1962.
19. Parzen, Emanuel, *Stochastic Processes*, Holden-Day, Inc., San Francisco, 1962.
20. Rainville, E. D., *Special Functions*, The Macmillan Company, New York, 1963.
21. Tucker, H. G., *An Introduction to Probability and Mathematical Statistics*, Academic Press, Inc., New York, 1965.

Appendix

TABLE I
Cumulative Normal Distribution
$X \sim N(0,1)$

X	$F_X(x)$	$1 - F_X(x)$	$f_X(x)$	X	$F_X(x)$	$1 - F_X(x)$	$f_X(x)$
.00	.5000	.5000	.3989	1.80	.9641	.0359	.0790
.05	.5199	.4801	.3984	1.85	.9678	.0322	.0721
.10	.5398	.4602	.3970	1.90	.9713	.0287	.0656
.15	.5596	.4404	.3945	1.95	.9744	.0256	.0596
.20	.5793	.4207	.3910	2.00	.9773	.0227	.0540
.25	.5987	.4013	.3867	2.05	.9798	.0202	.0488
.30	.6179	.3821	.3814	2.10	.9821	.0179	.0440
.35	.6368	.3632	.3752	2.15	.9842	.0158	.0396
.40	.6554	.3446	.3683	2.20	.9861	.0139	.0355
.45	.6736	.3264	.3605	2.25	.9878	.0122	.0317
.50	.6915	.3085	.3521	2.30	.9893	.0107	.0283
.55	.7088	.2918	.3429	2.35	.9906	.0094	.0252
.60	.7257	.2743	.3332	2.40	.9918	.0082	.0224
.65	.7422	.2578	.3230	2.45	.9929	.0071	.0198
.70	.7580	.2420	.3123	2.50	.9938	.0062	.0175
.75	.7734	.2266	.3011	2.55	.9946	.0054	.0155
.80	.7881	.2119	.2897	2.60	.9953	.0047	.0136
.85	.8023	.1977	.2870	2.65	.9960	.0040	.0119
.90	.8159	.1841	.2661	2.70	.9965	.0035	.0104
.95	.8289	.1711	.2541	2.75	.9970	.0030	.0091
1.00	.8413	.1587	.2420	2.80	.9974	.0026	.0079
1.05	.8531	.1469	.2299	2.85	.9978	.0022	.0069

1.10	.8643	.1357	.2179	2.90	.9981	.0019	.0060
1.15	.8749	.1251	.2059	2.95	.9984	.0016	.0051
1.20	.8849	.1151	.1942	3.00	.9987	.0013	.0044
1.25	.8944	.1056	.1826	3.05	.9989	.0011	.0038
1.30	.9032	.0968	.1714	3.10	.9990	.0010	.0033
1.35	.9115	.0885	.1604	3.15	.9992	.0008	.0028
1.40	.9192	.0808	.1497	3.20	.9993	.0007	.0024
1.45	.9265	.0735	.1394	3.25	.9994	.0006	.0020
1.50	.9332	.0668	.1295	3.30	.9995	.0005	.0017
1.55	.9394	.0606	.1200	3.35	.9996	.0004	.0015
1.60	.9452	.0548	.1109	3.40	.9997	.0003	.0012
1.65	.9505	.0495	.1023	3.45	.9997	.0003	.0010
1.70	.9554	.0446	.0940	3.50	.9998	.0002	.0009
1.75	.9959	.0401	.0863				

X	1.282	1.645	1.960	2.326	2.576	3.090
$F_X(x)$	.90	.95	.975	.99	.995	.999

TABLE II
Cumulative Chi-Square Distribution
Values of x such that for n degrees of freedom, $F_X(x) = P[X \leq x]$

n \ $F_X(x)$	.025	.05	.10	.20
2	.0506	.103	.211	.446
3	.216	.352	.584	1.00
4	.484	.711	1.06	1.65
5	.831	1.15	1.61	2.34
6	1.24	1.64	2.20	3.07
7	1.69	2.17	2.83	3.82
8	2.18	2.73	3.49	4.59
9	2.70	3.33	4.17	5.38
10	3.25	3.94	4.87	6.18
11	3.82	4.57	5.58	6.99
12	4.40	5.23	6.30	7.81
13	5.01	5.89	7.04	8.63
14	5.63	6.57	7.79	9.47
15	6.26	7.26	8.55	10.3
16	6.91	7.96	9.31	11.2
17	7.56	8.67	10.1	12.0
18	8.23	9.39	10.9	12.9
19	8.91	10.1	11.7	13.7
20	9.59	10.9	12.4	14.6

TABLE III

Cumulative *t*-Distribution

Values of x such that for n degrees of freedom, $F_X(x) = P[X \leq x]$

n \ $F_X(x)$	.75	.90	.95	.975	.99
1	1.000	3.078	6.314	12.706	31.821
2	.816	1.886	2.920	4.303	6.965
3	.765	1.638	2.353	3.182	4.541
4	.741	1.533	2.132	2.776	3.747
5	.727	1.476	2.015	2.571	3.365
6	.718	1.440	1.943	2.447	3.143
7	.711	1.415	1.895	2.365	2.998
8	.706	1.397	1.860	2.306	2.896
9	.703	1.383	1.833	2.262	2.821
10	.700	1.372	1.812	2.228	2.764
11	.697	1.363	1.796	2.201	2.718
12	.695	1.356	1.782	2.179	2.681
13	.694	1.350	1.771	2.160	2.650
14	.692	1.345	1.761	2.145	2.624
15	.691	1.341	1.753	2.131	2.602
16	.690	1.337	1.746	2.120	2.583
17	.689	1.333	1.740	2.110	2.567
18	.688	1.330	1.734	2.101	2.552
19	.688	1.328	1.729	2.093	2.539
20	.687	1.325	1.725	2.086	2.528
25	.684	1.316	1.708	2.060	2.485
30	.683	1.310	1.697	2.042	2.457
40	.681	1.303	1.684	2.021	2.423
∞	.674	1.282	1.645	1.960	2.326

TABLE IV

Cumulative *F*-Distribution

Values of x such that $F_X(x) = P[X \leq x]$ when $X \sim F(n, m)$

m	$F_X(x)$	1	2	3	4	5	6	7	8	9	10	11	12	15	19
1	.025	$.0^{2}15$*	.026	.057	.082	.100	.113	.124	.132	.139	.144	.149	.153	.161	.170
	.05	$.0^{2}62$	.054	.099	.130	.151	.167	.179	.188	.195	.201	.207	.211	.220	.230
	.10	.025	.117	.181	.220	.246	.265	.279	.289	.298	.304	.310	.315	.325	.336
	.90	39.9	49.5	53.6	55.8	57.2	58.2	58.9	59.4	59.9	60.2	60.5	60.7	61.2	61.7
	.95	161	200	216	225	230	234	237	239	241	242	243	244	246	248
	.975	648	800	864	900	922	937	948	957	963	969	973	977	985	993
2	.025	$.0^{2}13$	.026	.062	.094	.119	.138	.153	.165	.175	.183	.190	.196	.210	.224
	.05	$.0^{2}50$	.053	.105	.144	.173	.194	.211	.224	.235	.244	.251	.257	.272	.286
	.10	.020	.111	.183	.231	.265	.289	.307	.321	.333	.342	.350	.356	.371	.386
	.90	8.53	9.00	9.16	9.24	9.29	9.33	9.35	9.37	9.38	9.39	9.40	9.41	9.42	9.44
	.95	18.5	19.0	19.2	19.2	19.3	19.3	19.4	19.4	19.4	19.4	19.4	19.4	19.4	19.4
	.975	38.5	39.0	39.2	39.2	39.3	39.3	39.4	39.4	39.4	39.4	39.4	39.4	39.4	39.4
3	.025	$.0^{2}12$	.026	.065	.100	.129	.152	.170	.185	.197	.207	.216	.224	.241	.259
	.05	$.0^{2}46$	.052	.108	.152	.185	.210	.230	.246	.259	.270	.279	.287	.304	.323
	.10	.019	.109	.185	.239	.276	.304	.325	.342	.356	.367	.376	.384	.402	.420
	.90	5.54	5.46	5.39	5.34	5.31	5.28	5.27	5.25	5.24	5.23	5.22	5.22	5.20	5.18
	.95	10.1	9.55	9.28	9.12	9.01	8.94	8.89	8.85	8.81	8.79	8.76	8.74	8.70	8.66
	.975	17.4	16.0	15.4	15.1	14.9	14.7	14.6	14.5	14.5	14.4	14.4	14.3	14.3	14.2

4	.025	.011	.026	.066	.104	.135	.161	.181	.198	.212	.224	.234	.243	.262	.284
	.05	.044	.052	.110	.157	.193	.221	.243	.261	.275	.288	.298	.307	.327	.349
	.10	.018	.108	.187	.243	.284	.314	.338	.356	.371	.384	.394	.403	.424	.445
	.90	4.54	4.32	4.19	4.11	4.05	4.01	3.98	3.95	3.94	3.92	3.91	3.90	3.87	3.84
	.95	7.71	6.94	6.59	6.39	6.26	6.16	6.09	6.04	6.00	5.96	5.94	5.91	5.86	5.80
	.975	12.2	10.6	9.98	9.60	9.36	9.20	9.07	8.98	8.90	8.84	8.79	8.75	8.66	8.56
5	.025	.011	.025	.067	.107	.140	.167	.189	.208	.223	.236	.248	.257	.280	.304
	.05	.043	.052	.111	.160	.198	.228	.252	.271	.287	.301	.313	.322	.345	.369
	.10	.017	.108	.188	.247	.290	.322	.347	.367	.383	.397	.408	.418	.440	.463
	.90	4.06	3.78	3.62	3.52	3.45	3.40	3.37	3.34	3.32	3.30	3.28	3.27	3.24	3.21
	.95	6.61	5.79	5.41	5.19	5.05	4.95	4.88	4.82	4.77	4.74	4.71	4.68	4.62	4.56
	.975	10.0	8.43	7.76	7.39	7.15	6.98	6.85	6.76	6.68	6.62	6.57	6.52	6.43	6.33
6	.025	.011	.025	.068	.109	1.43	1.72	.195	.215	.231	.246	.258	.268	.293	.320
	.05	.043	.052	.112	.162	.202	.233	.259	.279	.296	.311	.324	.334	.358	.385
	.10	.017	.107	.189	.249	.294	.327	.354	.375	.392	.406	.418	.429	.453	.478
	.90	3.78	3.46	3.29	3.18	3.11	3.05	3.01	2.98	2.96	2.94	2.92	2.90	2.87	2.84
	.95	5.99	5.14	4.76	4.53	4.39	4.28	4.21	4.15	4.10	4.06	4.03	4.00	3.94	3.87
	.975	8.81	7.26	6.60	6.23	5.99	5.82	5.70	5.60	5.52	5.46	5.41	5.37	5.27	5.17
7	.025	$.0^{2}10$	.025	.068	.110	.146	.176	.200	.221	.238	.253	.266	.277	.304	.333
	.05	$.0^{2}42$	.052	.113	.164	.205	.238	.264	.286	.304	.319	.332	.343	.369	.398
	.10	.017	.107	.190	.251	.297	.332	.359	.381	.399	.414	.427	.438	.463	.491
	.90	3.59	3.26	3.07	2.96	2.88	2.83	2.78	2.75	2.72	2.70	2.68	2.67	2.63	2.59
	.95	5.59	4.74	4.35	4.12	3.97	3.87	3.79	3.73	3.68	3.64	3.60	3.57	3.51	3.44
	.975	8.07	6.54	5.89	5.52	5.29	5.12	4.99	4.90	4.82	4.76	4.71	4.67	4.57	4.47

TABLE IV (Continued)
Cumulative *F*-Distribution
Values of x such that $F_X(x) = P[X \leq x]$ when $X \sim F(n, m)$

m	$F_X(x)$	1	2	3	4	5	6	7	8	9	10	11	12	15	19
8	.025	$.0^2 10$	.025	.069	.111	.148	.179	.204	.226	.244	.259	.273	.285	.313	.343
	.05	$.0^2 42$	.052	.113	.166	.208	.241	.268	.291	.310	.326	.339	.351	.379	.409
	.10	.017	.107	.190	.253	.299	.335	.363	.386	.405	.421	.435	.445	.472	.500
	.90	3.46	3.11	2.92	2.81	2.73	2.67	2.62	2.59	2.56	2.54	2.52	2.50	2.46	2.42
	.95	5.32	4.46	4.07	3.84	3.69	3.58	3.50	3.44	3.39	3.35	3.31	3.28	3.22	3.15
	.975	7.57	6.06	5.42	5.05	4.82	4.65	4.53	4.43	4.36	4.30	4.24	4.20	4.10	4.00
9	.025	$.0^2 10$	.025	.069	.112	.150	.181	.207	.230	.248	.265	.279	.291	.320	.352
	.05	$.0^2 40$	.052	.113	.167	.210	.244	.272	.296	.315	.331	.345	.358	.386	.418
	.10	.017	.107	.191	.254	.302	.338	.367	.390	.410	.426	.441	.452	.479	.509
	.90	3.36	3.01	2.81	2.69	2.61	2.55	2.51	2.47	2.44	2.42	2.40	2.38	2.34	2.30
	.95	5.12	4.26	3.86	3.63	3.48	3.37	3.29	3.23	3.18	3.14	3.10	3.07	3.01	2.94
	.975	7.21	5.71	5.08	4.72	4.48	4.32	4.20	4.10	4.03	3.96	3.91	3.87	3.77	3.67
10	.025	$.0^2 10$	.025	.069	.113	.151	.183	.210	.233	.252	.269	.283	.296	.327	.360
	.05	$.0^2 41$	.052	.114	.168	.211	.246	.275	.299	.319	.336	.351	.363	.393	.426
	.10	.017	.106	.191	.255	.303	.340	.370	.394	.414	.430	.444	.457	.486	.516
	.90	3.28	2.92	2.73	2.61	2.52	2.46	2.41	2.38	2.35	2.32	2.30	2.28	2.24	2.20
	.95	4.96	4.10	3.71	3.48	3.33	3.22	3.14	3.07	3.02	2.98	2.94	2.91	2.85	2.77
	.975	6.94	5.46	4.83	4.47	4.24	4.07	3.95	3.85	3.70	3.72	3.66	3.62	3.52	3.42

11	.025	$.0^{2}10$	.025	.069	.114	.152	.185	.212	.236	.256	.273	.288	.301	.332	.368
	.05	$.0^{2}41$	.052	.114	.168	.212	.248	.278	.302	.323	.340	.355	.368	.398	.433
	.10	.017	.106	.192	.256	.305	.342	.373	.397	.417	.435	.448	.461	.490	.524
	.90	3.23	2.86	2.66	2.54	2.45	2.39	2.34	2.30	2.27	2.25	2.23	2.21	2.17	2.12
	.95	4.84	3.98	3.59	3.36	3.20	3.09	3.01	2.95	2.90	2.85	2.82	2.79	2.72	2.65
	.975	6.72	5.26	4.63	4.28	4.04	3.88	3.76	3.66	3.59	3.53	3.47	3.43	3.33	3.23
12	.025	$.0^{2}10$	.025	.070	.114	.153	.186	.214	.238	.259	.276	.292	.305	.337	.374
	.05	$.0^{2}41$	.052	.114	.169	.214	.250	.280	.305	.325	.343	.358	.372	.404	.439
	.10	.016	.106	.192	.257	.306	.344	.375	.400	.420	.438	.452	4.66	.496	.528
	.90	3.18	2.81	2.61	2.48	2.39	2.33	2.28	2.24	2.21	2.19	2.17	2.15	2.11	2.06
	.95	4.75	3.89	3.49	3.26	3.11	3.00	2.91	2.85	2.80	2.75	2.72	2.69	2.62	2.54
	.975	6.55	5.10	4.47	4.12	3.89	3.73'	3.61	3.51	3.44	3.37	3.32	3.28	3.18	3.07

*The notation $.0^{2}15$ reads .0015, $.0^{3}16$ reads .00016, etc.

Symbols

$E[X]$	expected value of X
$F(m, n)$	F distribution with m and n degrees of freedom
F_X	probability distribution function of the random variable X
F_{XY}	joint probability distribution function of the random variables X and Y
$F_{X\mid Y}$	conditional probability distribution function of X given Y
f_X	probability density function of the random variable X
f_{XY}	joint probability density function of the random variables X and Y
$f_{X\mid Y}$	conditional probability density function of X given Y
$L(\chi; \mu)$	likelihood function at the points $\chi = (x_1, x_2, \ldots, x_n)$ and μ
$M_X(t)$	moment generating function of the random variable X at the point t
$N!$	product $N(N-1)(N-2)\cdots 1$
$N(\mu, \sigma^2)$	normal distribution with mean μ and variance σ^2
P	probability function
$P[E]$	probability of the event E
$P[E_1 \mid E_2]$	probability of the event E_1 given the event E_2
$P[X \leq x]$	probability that the random variable X has a value equal to or less than the real number x
pdf	probability density function
$R_X(\tau)$	autocorrelation of X at τ
$R_{XY}(\tau)$	cross correlation of X and Y at τ
$S_X(\omega)$	spectral density or power spectrum of X at the frequency ω
$S_{XY}(\omega)$	cross spectrum of X and Y at the frequency ω

$t(n)$	t distribution with n degrees of freedom
X	random variable
$X \sim f_X$	X has the probability density function f_X
x	value of the random variable X
$x \in A$	x is an element of the set A
Γ	gamma function
σ_X^2	variance of X
$\chi^2(n)$	chi-square distribution with n degrees of freedom

Index